Optimal Domain and Integral Extension of Operators Acting in Fréchet Function Spaces

Bettina Blaimer

Das vorliegende Werk wurde im Sommersemester 2017 an der
Mathematisch-Geographischen Fakultät der Katholischen Universität
Eichstätt-Ingolstadt unter gleichnamigen Titel als Dissertation
im Fach Mathematik angenommen.

Bibliografische Information der Deutschen Nationalbibliothek

Die Deutsche Nationalbibliothek verzeichnet diese Publikation in der
Deutschen Nationalbibliografie; detaillierte bibliografische Daten sind
im Internet über http://dnb.d-nb.de abrufbar.

ISBN 978-3-8325-4557-4

Logos Verlag Berlin GmbH
Comeniushof, Gubener Str. 47,
10243 Berlin
Tel.: +49 (0)30 42 85 10 90
Fax: +49 (0)30 42 85 10 92
INTERNET: http://www.logos-verlag.de

Contents

Chapter 1

Introduction

The idea underlying this thesis is based on a problem which is, for Banach function spaces, fully developed and may at the beginning be called the "optimal domain problem". To describe this term more precisely let us start with a simpler version of this problem. Suppose that $X(\mu)$ is a Banach function space over a positive, finite measure space (Ω, Σ, μ), X is a Banach space and T is a continuous linear operator defined on $X(\mu)$ and assuming its values in X. Is it then possible to find a larger Banach function space $Y(\mu)$ over (Ω, Σ, μ) including $X(\mu)$ continuously and such that T extended to $Y(\mu)$ is still continuous?

The answer is "yes" as the following example shows. Consider the Volterra operator $V_1 : L^1([0,1]) \to L^1([0,1])$ mapping $f \mapsto V_1(f)$ where

$$V_1(f)(w) := \int_0^w f(t)\, d\lambda(t), \quad \text{for } w \in [0,1].$$

Hence, V_1 is defined on the Banach function space $L^1([0,1])$ over the positive, finite measure space $\big([0,1], \mathcal{B}([0,1]), \lambda\big)$ where λ denotes Lebesgue measure and $\mathcal{B}([0,1])$ the Lebesgue measurable subsets of $[0,1]$. Obviously we can choose

$$f(t) := (1-t)^{-1}, \quad \text{for } t \in [0,1),$$

to obtain a function which is certainly not in $L^1([0,1])$ but nevertheless suggests a candidate for an element of an extended domain of V_1 since

$$\int_0^w (1-t)^{-1}\, d\lambda(t) = -\big(\ln(1-w)\big), \quad \text{for all } w \in [0,1),$$

and, moreover, $-\ln(1 - \cdot) \in L^1([0,1])$ as direct calculation shows. Indeed, the investigations in [28] revealed that the original domain of V_1 can be extended to the larger space $L^1\big((1-t)\, d\lambda(t)\big)$.

This result inevitably leads to the question whether there exists a sort of "largest domain" of T. More precisely: Let $X(\mu)$ be a Banach function space over a positive, finite measure space (Ω, Σ, μ), X be a Banach space and $T : X(\mu) \to X$ be a continuous linear operator again. Does there exist a largest Banach function space $Z(\mu)$ over (Ω, Σ, μ) satisfying $X(\mu) \subseteq Z(\mu)$ continuously (in the sense that $Y(\mu) \subseteq Z(\mu)$ for all Banach function spaces $Y(\mu)$ into which $X(\mu)$ is included continuously) and such there is a continuous linear operator $\tilde{T} : Z(\mu) \to X$ which coincides with T on $X(\mu)$? This question was first successfully and systematically treated for various kernel operators T by Curbera and Ricker; see, for example, the papers [3], [4].

One can show that under certain assumptions on the space $X(\mu)$ the existence of such a "largest" Banach function space follows immediately. But the amazing point is that under these assumptions on $X(\mu)$ there is a remarkable connection between the search of the optimal domain of T and the theory of vector measures. So, let $X(\mu)$ be a Banach function space over a positive, finite measure space (Ω, Σ, μ) again, this time, however, containing the Σ-simple functions $\mathrm{sim}(\Sigma)$ and such that its norm $\|\cdot\|_{X(\mu)}$ is σ-order continuous. Then, $\mathrm{sim}(\Sigma)$ is necessarily dense in $X(\mu)$. Moreover, let X be a Banach space and $T : X(\mu) \to X$ be a continuous linear operator. The finitely additive set function $m_T : \Sigma \to X$ defined by

$$m_T(A) := T(\chi_A), \quad \text{for } A \in \Sigma,$$

is then a Banach-space-valued vector measure and, for each $s \in \mathrm{sim}(\Sigma)$, the equation

$$\int_\Omega s \, dm_T = T(s)$$

holds. Whenever additionally the m_T-null sets and the μ-null sets coincide, one obtains the following result: The optimal domain space of T exists and coincides with the space $L^1(m_T)$ of all m_T-integrable functions. Furthermore, the optimal extension of T is the integration operator $I_{m_T} : L^1(m_T) \to X$ given by

$$I_{m_T}(f) := \int_\Omega f \, dm_T, \quad \text{for } f \in L^1(m_T).$$

The so-called "optimal extension process" of continuous linear operators defined on Banach function spaces as discussed in the previous paragraph has been studied thoroughly by various mathematicians. Moreover, the search for and characterization of the optimal domain resp. the optimal extension of T has found a variety of applications, for example, the study of kernel and differential operators, to name but a few. A large part of the current research on this topic is summarized in [26];

6

the results established therein will form the foundations of this thesis.

Since not all of the important spaces in analysis are normable it is only natural to ask whether this "optimal extension process" can also be applied to other spaces than only Banach function spaces. So, the aim of this thesis will be to translate the theory to a special class of function spaces, namely the Fréchet function spaces, whose topology is not generated by a single function norm but by a sequence of function semi-norms. Starting-point will then be a Fréchet function space $X(\mu)$ over a positive, σ-finite measure space (Ω, Σ, μ) and a continuous linear operator $T : X(\mu) \to X$ with values in a Fréchet space X. Hence, we are looking for a "largest" Fréchet function space $Z(\mu)$ including $X(\mu)$ continuously (in the sense as stated above) and such that the extension $\tilde{T} : Z(\mu) \to X$ is still continuous and coincides with T on $X(\mu)$. The main goal will be to see whether the topology of the Fréchet function space $X(\mu)$ allows similar properties and concepts as in the case of Banach function spaces and thus, gives way to a connection between the search of the optimal domain and extension of T and the theory of Fréchet-space-valued vector measures. Of course, the central question is whether in the case of Fréchet function spaces the optimal domain of T is still $L^1(m_T)$ and its optimal domain the integration operator I_{m_T}.

The thesis is structured as follows.

Chapter 2 provides all the necessary mathematical foundations of this work. Not only notations will be fixed, but also all the lemmas and propositions that turn out to be relevant for the theory will be formulated or, if necessary, be derived in this chapter.

Chapter 3 forms the theoretical part of this thesis. At first we will take a closer look at the Fréchet function space $X(\mu)$ and its properties. Special attention will be paid to the inclusion $X(\mu) \subseteq Y(\mu)$ for any Fréchet function space $Y(\mu)$ containing $X(\mu)$. In a second step we will concentrate on the Fréchet-space-valued vector measure m_T associated with T and the space of m_T-integrable functions $L^1(m_T)$. Finally, we will decide whether the optimal domain of T exists and coincides with $L^1(m_T)$.

Chapter 4 intends to apply the theory obtained in Chapter 3 to some well-known operators $T : X(\mu) \to X$ defined on Fréchet function spaces $X(\mu)$, namely the multiplication operators $M_g^{p-} : L^{p-}([0,1]) \to L^{p-}([0,1])$ and $M_{g,\mathrm{loc}}^p : L_{\mathrm{loc}}^p(\mathbb{R}) \to L_{\mathrm{loc}}^p(\mathbb{R})$, the Volterra operator $V_{p-} : L^{p-}([0,1]) \to L^{p-}([0,1])$ and the convolution operator $C_g^{p-} : L^{p-}(G) \to L^{p-}(G)$ (where G is a compact Abelian group).

Chapter 5 summarizes the results obtained in Chapter 3 and Chapter 4 and will give a preview of possible further research.

Chapter 2

Preliminaries

Chapter 2 presents an overview over all the mathematical fields that are relevant
for this thesis. The aim is to introduce the definitions and notations that will be
used in the forthcoming chapters and to highlight some well-known (and some not
so well-known) results that will be relevant especially for the theoretical part of this
work. Section 2.1 treats the theory of locally convex topological vector spaces and
the different topologies defined on such a space. Special attention will be paid to
the class of Fréchet spaces and their properties since they will play a major role in
the sequel. Section 2.2 summarizes classical measure and integration theory with
regard to a finite, σ-finite or complex measure μ. Here we will concentrate mainly
on the connection between convergence μ-a.e. and local convergence in measure as
well as on the topology of local convergence in measure. Section 2.3 deals with a
special class of Riesz spaces, the Fréchet function spaces. The focus will be on the
properties of the Fréchet function spaces and their relevance for the forthcoming
studies. The section closes with two examples of Fréchet function spaces that will
play a significant role in Chapter 4: the spaces $L^{p-}([0,1])$ and $L^p_{\mathrm{loc}}(\mathbb{R})$. Section
2.4 introduces the terminology concerning vector measures having values in Fréchet
spaces. Since it will turn out to be a useful tool for the applications in Chapter 4
we take a look at the Bochner μ-integrability and the Pettis μ-integrability as well.
Finally, Section 2.5 outlines the theory of integration on topological groups. It will
become important when the convolution operator is investigated in Section 4.3.

2.1   Fréchet spaces

Let X be a vector space over a scalar field $\mathbb{K}$ (where $\mathbb{K} = \mathbb{R}$ or $\mathbb{K} = \mathbb{C}$) and denote
by $\mathcal{P}(X)$ the set of all subsets of X. Let $\tau \subseteq \mathcal{P}(X)$ be a system of subsets of X.
Recall that τ is called a *topology* on X if it satisfies the following conditions:

(i) $X \in \tau, \varnothing \in \tau$.

(ii) If $V, W \in \tau$, then also $V \cap W \in \tau$.

(iii) If $\{V_j\}_{j \in J} \subseteq \tau$, then also $\bigcup_{j \in J} V_j \in \tau$.

The elements of τ are called the open sets of X whereas their complements are the closed sets of X. Note, given two vector spaces X_1, X_2 equipped with a topology τ_1 resp. τ_2, that an operator $T : X_1 \to X_2$ is called *continuous* if $T^{-1}(V) := \{x \in X_1 : T(x) \in V\} \in \tau_1$, for all $V \in \tau_2$. A vector space X endowed with a topology τ such that the vector space operations are continuous with respect to τ is called a *topological vector space*. It is denoted by the pair (X, τ).

Let $V \in \mathcal{P}(X)$ be a subset of a topological vector space (X, τ). V is called *balanced* if, for every $x \in V$ and for every $\lambda \in \mathbb{K}$ satisfying $|\lambda| \leqslant 1$, the element λx is in V again. V is said to be *absorbing* if $X = \bigcup_{n \in \mathbb{N}} nV$, where $nV = \{nx : x \in V\}$. It is called *convex* if, for any $x, y \in V$, the line segment $\lambda x + (1 - \lambda)y$, where $0 \leqslant \lambda \leqslant 1$, is contained in V. The set V is said to be *absolutely convex* if it is balanced and convex. Finally, V is called *compact* if, for every collection $\{V_j\}_{j \in J} \subseteq \tau$ (J being an arbitrary index set) satisfying $V \subseteq \bigcup_{j \in J} V_j$, there exists a finite subset $F \subseteq J$ such that $V \subseteq \bigcup_{j \in F} V_j$.

Let (X, τ) be a topological vector space and let $x \in X$. A subset $U \in \mathcal{P}(X)$ is called a *neighbourhood* of x if there is a set $V \in \tau$ satisfying $x \in V$ and $V \subseteq U$. The topological vector space X is called *Hausdorff* if, for each pair $x, y \in X$ of distinct points, there are respective neighbourhoods U_x, U_y of x, y such that $U_x \cap U_y = \varnothing$. The set of all neighbourhoods of x is denoted by $\mathcal{U}(x)$. A subset $\mathcal{B} \subseteq \mathcal{U}(x)$ is called a *neighbourhood base* if, for each $U \in \mathcal{U}(x)$, there is a set $B \in \mathcal{B}$ such that $B \subseteq U$. The topological vector space X is called a *locally convex topological vector space* if each element of X has a neighbourhood base of absolutely convex sets. A system $\mathcal{V}$ of neighbourhoods of 0 is called *fundamental* if, for each neighbourhood $U \in \mathcal{U}(0)$, there exists $\tilde{U} \in \mathcal{V}$ such that $\tilde{U} \subseteq U$.

Recall that a mapping $p : X \to [0, \infty)$ on a vector space X is called a *semi-norm* if it satisfies the following conditions:

(i) If $x = 0$, then $p(x) = 0$.

(ii) $p(\lambda x) = |\lambda| \, p(x)$, for all $\lambda \in \mathbb{K}$, for all $x \in X$.

(iii) $p(x + y) \leqslant p(x) + p(y)$, for all $x, y \in X$.

A semi-norm p is called a *norm* if condition (i) is valid in both directions:

(i') $x = 0$ if and only if $p(x) = 0$.

Let X be a locally convex topological vector space and let $U \subseteq X$ be an absolutely convex, absorbing set. The *Minkowski functional* $\phi_U : X \to [0, \infty)$ is defined by

$$\phi_U(x) := \inf\{\lambda > 0 : x \in \lambda U\}, \quad \text{for } x \in X. \tag{2.1}$$

In the sequel, we will consider a family $\{p_i\}_{i \in I}$ (with I an arbitrary index set) of continuous semi-norms on a topological vector space X. $\{p_i\}_{i \in I}$ is said to be a *fundamental system of semi-norms* if the sets $U_i := \{x \in X : p_i(x) < 1\}$, where $i \in I$, form a fundamental system of neighbourhoods of 0. Note that every locally convex topological vector space has a fundamental system of semi-norms $\{p_i\}_{i \in I}$ meaning that $\{p_i\}_{i \in I}$ satisfies the following conditions, [22, p. 232]:

(i) For each $x \neq 0$, there is an $i \in I$ such that $p_i(x) > 0$.
(ii) For any $i, j \in I$, there exist $a \in I$ and $M > 0$ such that $\max\{p_i, p_j\} \leqslant M\, p_a$.

A family of semi-norms which satisfies condition (i) is called *separated*. A locally convex topological vector space is *Hausdorff* if and only if it has a separated family of continuous semi-norms. Conversely, let X be a vector space and let $\{p_i\}_{i \in I}$ be any family of semi-norms satisfying conditions (i) and (ii) as stated before. Then there is a unique locally convex topology on X such that the $\{p_i\}_{i \in I}$ form a fundamental system of semi-norms [22, p. 233]. A locally convex topological vector space X whose topology is generated by a fundamental system of semi-norms $\{p_i\}_{i \in I}$ will be denoted by the pair $(X, \{p_i\}_{i \in I})$.

Let $(X, \{p_i\}_{i \in I})$ be a locally convex Hausdorff space and let $\{x_\delta\}_{\delta \in D} \subseteq X$ be a net in X (here, $(D, \geqslant)$ denotes a directed set). $\{x_\delta\}_{\delta \in D}$ is said to *converge* to an element $x \in X$ if

$$\lim_\delta p_i(x - x_\delta) = 0, \quad \text{for all } i \in I.$$

$\{x_\delta\}_{\delta \in D}$ is called *Cauchy* if, for every semi-norm p_i and for every $\varepsilon > 0$, there exists an index $\vartheta_{i,\varepsilon}$ such that

$$p_i(x_\delta - x_{\tilde\delta}) < \varepsilon, \quad \text{for all } \delta, \tilde\delta \geqslant \vartheta_{i,\varepsilon}.$$

Accordingly, a sequence $\{x_n\}_{n \in \mathbb{N}} \subseteq X$ is called *Cauchy* if, for every semi-norm p_i and for every $\varepsilon > 0$, there exists an index $N_{i,\varepsilon} \in \mathbb{N}$ such that

$$p_i(x_n - x_m) < \varepsilon, \quad \text{for all } n, m \geqslant N_{i,\varepsilon}.$$

In a locally convex Hausdorff space X we therefore have the following notations: X is said to be *complete* if every Cauchy net in X converges to some element in X, and X is called *sequentially complete* if every Cauchy sequence converges to some element in X. A locally convex Hausdorff space is called a *Fréchet space* if the topology of X is generated by a fundamental sequence of semi-norms $\{p_k\}_{k\in\mathbb{N}}$ such that X is complete. Note that in a Fréchet space completeness and sequential completeness coincide, [22, p. 239]. In the sequel, we assume a Fréchet space is always generated by an increasing fundamental system of semi-norms $\{p_k\}_{k\in\mathbb{N}}$, that is, $p_k \leqslant p_{k+1}$, for all $k \in \mathbb{N}$.

Recall that a mapping $d : X \times X \to [0, \infty)$ on a topological vector space X is called a *pseudo-metric* if the following conditions are satisfied:

(i) $d(x,x) = 0$.

(ii) $d(x,y) = d(y,x)$, for all $x, y \in X$.

(iii) $d(x,z) \leqslant d(x,y) + d(y,z)$, for all $x, y, z \in X$.

Note that points in a pseudo-metric space need not be distinguishable, that is, one may have $d(x,y) = 0$ for distinct values $x \neq y$. Whenever condition (i) can be replaced by

(i') $d(x,y) = 0$ if and only if $x = y$,

the mapping d is called a *metric*. Since a Fréchet space $(X, \{p_k\}_{k\in\mathbb{N}})$ has a countable fundamental system of semi-norms, X is *metrizable* means that there exists a metric d on X such that (X, d) is a metric space whose metric induces the topology on X and such that $(X, \{p_k\}_{k\in\mathbb{N}})$ and (X, d) have the same Cauchy sequences, [22, pp. 276–277]. A metric which satisfies the required conditions is defined by means of the semi-norms p_k as follows:

$$d(x,y) := \sum_{k=1}^{\infty} \frac{p_k(x-y)}{2^k\left(1 + p_k(x-y)\right)}, \quad \text{for } x, y \in X.$$

Let $(X, \{p_k\}_{k\in\mathbb{N}}), (Y, \{\tilde{p}_k\}_{k\in\mathbb{N}})$ be two Fréchet spaces and let $T : X \to Y$ be a linear operator. Then T is *continuous* if, for every $k \in \mathbb{N}$, there exist $l_k \in \mathbb{N}$ and $M_k > 0$, such that

$$\tilde{p}_k\left(T(x)\right) \leqslant M_k\, p_{l_k}(x), \quad \text{for all } x \in X. \tag{2.2}$$

A criterion for continuity by means of sequences is given by the following important theorem; see [18, p. 168] where it is stated for the more general case that X, Y are complete metrizable topological vector spaces.

Proposition 2.1.1 (Closed Graph Theorem)

Let X, Y be two Fréchet spaces and let $T : X \to Y$ be a linear operator. Then the following equivalence holds: T is continuous if and only if whenever $\lim_{n\to\infty} x_n = x$ in

X and $\lim_{n\to\infty} T(x_n) = y$ in Y, then $T(x) = y$. $\quad\square$

Note that a linear operator T from a Fréchet space X to a Fréchet space Y is continuous if and only if T is continuous in 0, [22, pp. 233–234]. Hence, the Closed Graph Theorem can be reduced to the following assertion.

Corollary 2.1.1

Let X, Y be two Fréchet spaces and let $T : X \to Y$ be a linear operator. Then the following equivalence holds: T is continuous if and only if whenever $\lim_{n\to\infty} x_n = 0$ in X and $\lim_{n\to\infty} T(x_n) = y$ in Y, then $y = 0$. $\quad\square$

In a Fréchet space $(X, \{p_k\}_{k\in\mathbb{N}})$ a linear functional $x^* : X \to \mathbb{C}$ is *continuous* if and only if there exist an index $k \in \mathbb{N}$ and a constant $M > 0$ such that

$$|\langle x, x^* \rangle| \leqslant M\, p_k(x), \quad \text{for all } x \in X.$$

Here, $\langle \cdot, \cdot \rangle$ denotes the canonical bilinear form of the duality. The above definition is equivalent to the assertion that there exists a neighbourhood $U \in \mathcal{U}(0)$ in X such that $\sup\{|\langle x, x^* \rangle| : x \in U\} < \infty$, [22, p. 234]. Denote by X^* the vector space of all such continuous linear functionals. X^* is called the *dual space* of X.

For each $x^* \in X^*$ define a semi-norm $p_{x^*} : X \to [0, \infty)$ by

$$p_{x^*}(x) := |\langle x, x^* \rangle|.$$

The family of semi-norms $\{p_{x^*}\}_{x^*\in X^*}$ induces the *weak topology* on X. It is the topology on X with the fewest open sets such that each element of X^* remains a continuous function with respect to the original topology on X. It is denoted by $\sigma(X, X^*)$. Moreover, the weak topology is locally convex, [22, p. 245]. A sequence $\{x_n\}_{n\in\mathbb{N}} \subseteq X$ is said to converge for the weak topology or *weakly converges* to $x \in X$ if and only if

$$\lim_{n\to\infty} \langle x_n, x^* \rangle = \langle x, x^* \rangle, \quad \text{for all } x^* \in X^*.$$

A subset $C \subseteq X$ is called *bounded* if

$$\sup\{p_k(x) : x \in C\} < \infty, \quad \text{for all } k \in \mathbb{N}.$$

A set $C \subseteq X$ is bounded in $\sigma(X, X^*)$ if

$$\sup\{|\langle x, x^* \rangle| : x \in C\} < \infty, \quad \text{for all } x^* \in X^*.$$

A set $C \subseteq X$ is bounded for the given locally convex Hausdorff topology in $(X, \{p_k\}_{k \in \mathbb{N}})$ if and only if it is bounded for the weak topology $\sigma(X, X^*)$, [22, p. 249]. For $V \subseteq X$ being any convex set the closure of V formed in the original topology of X coincides with its closure taken in the weak topology $\sigma(X, X^*)$, [32, p. 65].

Let (X, τ) be a locally convex Hausdorff space. It is possible to define a topology on X^* by means of a family of subsets of X. Let $\mathcal{C}$ denote the family of all bounded sets in X. A topology on X^* is generated by the semi-norms of the form

$$|x^*|_C := \sup\{|\langle x, x^* \rangle| : x \in C\}$$

as C varies over $\mathcal{C}$. This topology is called the *strong topology* of X^* and coincides with the topology of uniform convergence on the bounded sets in X. The dual space equipped with the strong topology is denoted by X_β^*. Note that X_β^* is locally convex again. The space of all continuous linear functionals on X_β^* is denoted by X^{**}. Equipped with the strong topology, i.e., $X^{**} = (X_\beta^*)_\beta^*$, it is called the *bidual* of X. The original space (X, τ) is called *reflexive* if the bidual X^{**} is equal to X and if the topology of $(X_\beta^*)_\beta^*$ coincides with the original topology τ.

For the next proposition let $(X, \{p_k\}_{k \in \mathbb{N}})$ be a Fréchet space. Denote, for each $k \in \mathbb{N}$, by X_k the completion of the quotient normed space $X/p_k^{-1}(\{0\})$. Then each X_k equipped with the quotient norm $\tilde{p}_k$ is a Banach space, a so-called *local Banach space*. The following result, [22, pp. 282–283], states a useful criterion concerning the reflexivity of a Fréchet space.

Proposition 2.1.2
If the Fréchet space X has a fundamental system of semi-norms $\{p_k\}_{k \in \mathbb{N}}$ such that all local Banach spaces X_k, for $k \in \mathbb{N}$, are reflexive, then X is reflexive.   □

Concerning the convergence of a series of elements of a Fréchet space $(X, \{p_k\}_{k \in \mathbb{N}})$ we have the following notations. Let $\sum_{n=1}^{\infty} x_n$ be a formal series in X. The series is said to *converge* in X, if there exists an element $x \in X$ such that the sequence of partial sums $\{\sum_{j=1}^{n} x_j\}_{n \in \mathbb{N}}$ converges to x in the topology of X. The series is said to be *unconditionally convergent* if, for every bijection $\pi : \mathbb{N} \to \mathbb{N}$, the series $\sum_{n=1}^{\infty} x_{\pi(n)}$ converges in X. The series is called *subseries convergent* if, for every increasing sequence of natural numbers $\{n_j\}_{j \in \mathbb{N}} \subseteq \mathbb{N}$, the series $\sum_{j=1}^{\infty} x_{n_j}$ is convergent in X. And finally, the series is *absolutely convergent* if $\sum_{n=1}^{\infty} p_k(x_n) < \infty$, for every $k \in \mathbb{N}$.

The following powerful theorem will be important for the study of vector measures (see [21], for instance).

Proposition 2.1.3 (Orlicz-Pettis Theorem)

Let X be a Fréchet space and let $\sum_{n=1}^{\infty} x_n$ be a series in X. Then the following assertions hold:

(i) $\sum_{n=1}^{\infty} x_n$ is unconditionally convergent in X if and only if $\sum_{n=1}^{\infty} x_n$ is subseries convergent in X.

(ii) Whenever each subseries $\sum_{j=1}^{\infty} x_{n_j}$ of the original series converges in X for the weak topology $\sigma(X, X^*)$, then $\sum_{n=1}^{\infty} x_n$ is subseries convergent in the topology of X. $\square$

In a topological vector space X an absolutely convex subset $V \subseteq X$ is called a *Banach disc* if its linear hull

$$X_V := \bigcup_{\lambda > 0} \lambda V,$$

equipped with its Minkowski functional ϕ_V as defined in (2.1), is a Banach space. Moreover, the natural injection $X_V \subseteq X$ is then continuous, [32, p. 97]. Each absolutely convex, bounded and closed subset V of a locally convex space X is a Banach disc, [22, p. 249].

Finally, a Hausdorff topological space X is called *locally compact* if each element $x \in X$ possesses a compact neighbourhood. Let X be a locally convex space, Y be a Fréchet space and $T : X \to Y$ a continuous linear map. Then T is called *compact* if there is a neighbourhood $U \in \mathcal{U}(0)$ such that the closure of its range $\overline{T(U)}$ is compact in Y.

2.2   Measure and measure space

Let Ω be a non-empty set and let $\mathcal{P}(\Omega)$ be the set of all subsets of Ω. A family of subsets $\Sigma \subseteq \mathcal{P}(\Omega)$ is called a *σ-algebra* if it satisfies the following conditions:

(i) $\Omega \in \Sigma$.

(ii)   If $A \in \Sigma$, then also its complement $A^c \in \Sigma$.

(iii)   If $\{A_j\}_{j \in \mathbb{N}} \subseteq \Sigma$, then also $\bigcup_{j \in \mathbb{N}} A_j \in \Sigma$.

Property (iii) can be substituted by the following equivalent condition:

(iii')   If $\{A_j\}_{j \in \mathbb{N}} \subseteq \Sigma$, then also $\bigcap_{j \in \mathbb{N}} A_j \in \Sigma$.

In the case of $\Omega = X$ being a topological or metric space and $\mathcal{O}$ being the system of open subsets of X, the σ-algebra of Borel sets, or *Borel σ-algebra*, is defined as the smallest σ-algebra over X containing $\mathcal{O}$. It is denoted by $\mathcal{B}(X)$. The pair (Ω, Σ) is called a *measurable space*.

A set function $\mu : \Sigma \to [0, \infty]$ defined on a σ-algebra Σ is called a *measure* if it satisfies $\mu(\varnothing) = 0$ and

$$\mu\left(\bigcup_{j=1}^{\infty} A_j\right) = \sum_{j=1}^{\infty} \mu(A_j), \tag{2.3}$$

for any sequence $\{A_j\}_{j \in \mathbb{N}} \subseteq \Sigma$ of disjoint sets. Condition (2.3) is called σ-*additivity* or countable additivity. It is characterized in the following way: Let $\mu : \Sigma \to [0, \infty]$ be a finitely additive set function and let $\{A_j\}_{j \in \mathbb{N}} \subseteq \Sigma$ be any sequence of sets such that $\mu(A_1) < \infty$ and $A_j \downarrow_j \varnothing$. Then μ is a measure, i.e., σ-additive, if and only if $\mu(A_j) \downarrow_j 0$, [11, p. 32]. Here, $\downarrow$ and $\uparrow$ indicate that the sets are monotone decreasing resp. increasing in Ω. A measure μ is called *finite* if $\mu(\Omega) < \infty$. It is called σ-*finite* if and only if there exists a sequence $\{A_j\}_{j \in \mathbb{N}} \subseteq \Sigma$ such that $\mu(A_j) < \infty$, for each $j \in \mathbb{N}$, and $\bigcup_{j \in \mathbb{N}} A_j = \Omega$. The triple (Ω, Σ, μ) is called a *measure space*.

Given a measure $\mu : \Sigma \to [0, \infty]$, a set $A \subseteq \Omega$ is called a μ-*null set* if $A \in \Sigma$ and $\mu(A) = 0$. The family of μ-null sets will be denoted by $\mathcal{N}_0(\mu)$. A measure space (Ω, Σ, μ) is called *complete* if each subset of a μ-null set $A \in \Sigma$ belongs to Σ and, hence, is a μ-null set as well. As usual, properties that are valid everywhere on Ω except on a μ-null set are said to be valid μ-*almost everywhere* (briefly: μ-a.e.).

Let (Ω, Σ, μ) be a measure space. A function $f : \Omega \to \mathbb{C}$ is said to be μ-*measurable* if $f^{-1}(A) := \{w \in \Omega : f(w) \in A\} \in \Sigma$, for every $A \in \mathcal{B}(\mathbb{C})$. Measurability of a function can also be characterized by means of Σ-simple functions. A function $s : \Omega \to \mathbb{C}$ is called a Σ-*simple function* if its range consists of finitely many points $\alpha_1, \dots, \alpha_l \in \mathbb{C}$ and if

$$A_j := s^{-1}(\{\alpha_j\}) = \{w \in \Omega : s(w) = \alpha_j\} \in \Sigma,$$

for each $j = 1, \dots, l$. Hence, a Σ-simple function s can also be written in the form

$$s = \sum_{j=1}^{l} \alpha_j \, \chi_{A_j}, \quad \text{where} \quad \bigcup_{j=1}^{l} A_j = \Omega. \tag{2.4}$$

If the points $\alpha_1, \dots, \alpha_l$ are distinct, then the sets $s^{-1}(\{\alpha_j\})$, for $j = 1, \dots, l$, are pairwise disjoint. It is clear that each Σ-simple function is measurable. The space of all Σ-simple functions will be denoted by $\mathrm{sim}(\Sigma)$. A function $f : \Omega \to [0, \infty)$ is μ-measurable if and only if there exists a sequence $\{s_n\}_{n \in \mathbb{N}}$ of Σ-simple functions satisfying $0 \leqslant s_n \uparrow_n f$. Applying this to $f^+ := \max\{f, 0\}$ and $f^- := \max\{-f, 0\}$ one obtains that, for each μ-measurable function $f : \Omega \to \mathbb{R}$, there exists a sequence $\{s_n\}_{n \in \mathbb{N}}$ of Σ-simple functions with $|s_n| \leqslant |f|$, for $n \in \mathbb{N}$, con-

16

verging to f pointwise on Ω. Thus, a $\mathbb{C}$-valued function f is μ-measurable if $(\mathrm{Re}(f))^+, (\mathrm{Re}(f))^-, (\mathrm{Im}(f))^+, (\mathrm{Im}(f))^-$ are μ-measurable. Here, $\mathrm{Re}(f)$ denotes the real part and $\mathrm{Im}(f)$ the imaginary part of the function f. The space of all $\mathbb{C}$-valued μ-measurable functions on Ω is denoted by $M(\mu)$. Note that functions in $M(\mu)$ differing only on a μ-null set are identified. Whenever we consider the set of all individual μ-measurable functions $f : \Omega \to \mathbb{C}$ we will write $\mathcal{M}(\mu)$. By $M(\mu)^+$ we denote the subset of all functions $f \in M(\mu)$ that are $\mathbb{R}$-valued and non-negative μ-a.e., i.e., functions satisfying $f \geqslant 0$ μ-a.e. on Ω.

Let (Ω, Σ, μ) be a measure space. For any Σ-simple function $s : \Omega \to [0, \infty)$ given by (2.4) its μ-integral is defined by

$$\int_\Omega s \, d\mu := \sum_{j=1}^l \alpha_j \, \mu(A_j). \tag{2.5}$$

Note that $\int_\Omega s \, d\mu = \infty$ whenever $\mu(A_j) = \infty$ and $\alpha_j \neq 0$ for some j. Accordingly, the μ-*integral* of a measurable function $f : \Omega \to [0, \infty)$, which is approximated by a sequence of non-negative Σ-simple functions $\{s_n\}_{n \in \mathbb{N}}$, that is, $0 \leqslant s_n \uparrow_n f$, is defined by

$$\int_\Omega f \, d\mu := \lim_{n \to \infty} \int_\Omega s_n \, d\mu.$$

Note that this definition is independent of the choice of the sequence $\{s_n\}_{n \in \mathbb{N}} \subseteq \mathrm{sim}(\Sigma)$, [11, p. 122]. A μ-measurable function $f : \Omega \to [0, \infty)$ is called μ-*integrable* if and only if its μ-integral over Ω takes a finite value. Accordingly, a μ-measurable function $f : \Omega \to \mathbb{R}$ is defined to be μ-integrable, if the μ-integrals of both f^+ and f^- take finite values. The μ-integral of f over Ω is then defined by

$$\int_\Omega f \, d\mu = \int_\Omega f^+ \, d\mu - \int_\Omega f^- \, d\mu.$$

A μ-measurable $\mathbb{C}$-valued function $f : \Omega \to \mathbb{C}$ is called μ-integrable if and only if the μ-integrals of both $\mathrm{Re}(f)$ and $\mathrm{Im}(f)$ take finite values. The integral of f is then defined by

$$\int_\Omega f \, d\mu = \int_\Omega \mathrm{Re}(f) \, d\mu + i \int_\Omega \mathrm{Im}(f) \, d\mu.$$

Note that in the case of a $\mathbb{C}$-valued function we have, [11, p. 129],

$$\mathrm{Re} \left(\int_\Omega f \, d\mu \right) = \int_\Omega \mathrm{Re}(f) \, d\mu, \quad \mathrm{Im} \left(\int_\Omega f \, d\mu \right) = \int_\Omega \mathrm{Im}(f) \, d\mu. \tag{2.6}$$

Denote by $\mathcal{L}^1(\mu)$ the space of all μ-measurable functions $f : \Omega \to \mathbb{C}$ satisfying

$$\|f\|_1 := \int_\Omega |f| \, d\mu < \infty.$$

Since $\mathcal{L}^1(\mu)$ is not Hausdorff as soon as there is a non-empty μ-null set, define by $\mathcal{N}(\mu)$ the set of all μ-measurable functions that are μ-*null*, that is, $f = 0$ μ-a.e. on Ω. The quotient $L^1(\mu) := \mathcal{L}^1(\mu)/\mathcal{N}(\mu)$ becomes a Hausdorff space and $\|\cdot\|_1$ is a norm on $L^1(\mu)$. Similarly, define for a fixed $p \in (1,\infty)$ the space $\mathcal{L}^p(\mu)$ consisting of all p-integrable functions, i.e., μ-measurable functions $f : \Omega \to \mathbb{C}$ that satisfy

$$\|f\|_p := \left(\int_\Omega |f|^p \, d\mu \right)^{1/p} < \infty.$$

Forming the quotient $L^p(\mu) := \mathcal{L}^p(\mu)/\mathcal{N}(\mu)$ we again obtain a Hausdorff space and $\|\cdot\|_p$ becomes a norm on $L^p(\mu)$. Finally, denote by $\mathcal{L}^\infty(\mu)$ the space of all μ-measurable functions $f : \Omega \to \mathbb{C}$ satisfying

$$\|f\|_\infty := \operatorname{ess\,sup} \left\{ |f(w)| : w \in \Omega \right\} < \infty,$$

the so-called space of all μ-essentially bounded functions in $\mathcal{M}(\mu)$, i.e., functions that are bounded except on a μ-null set. Again, $L^\infty(\mu) := \mathcal{L}^\infty(\mu)/\mathcal{N}(\mu)$ is Hausdorff and $\|\cdot\|_\infty$ a norm on $L^\infty(\mu)$. For $1 \leqslant p \leqslant \infty$, each space $L^p(\mu)$ is a complete normed space, i.e., a Banach space, [11, p. 232]. Furthermore, in finite measure spaces the inclusions

$$L^\infty(\mu) \subseteq L^{p'}(\mu) \subseteq L^p(\mu) \subseteq L^1(\mu) \tag{2.7}$$

hold, for $1 < p < p' < \infty$. Concerning the norms we have

$$\|f\|_p \leqslant \mu(\Omega)^{(1/p)-(1/p')} \|f\|_{p'}, \tag{2.8}$$

for all $f \in L^{p'}(\mu)$, [11, p. 233]. The inequality (2.8) results from *Hölder's inequality*, i.e.,

$$\int_\Omega |fg| \, d\mu \leqslant \left(\int_\Omega |f|^p \, d\mu \right)^{1/p} \left(\int_\Omega |g|^q \, d\mu \right)^{1/q} \tag{2.9}$$

which is valid for all μ-measurable functions $f, g : \Omega \to \mathbb{C}$ and $1 \leqslant p, q \leqslant \infty$ satisfying $\frac{1}{p} + \frac{1}{q} = 1$ where $\frac{1}{\infty} := 0$, [11, p. 223]. For $p \in (1,\infty)$ fixed, let q be the *conjugate exponent* of p meaning that $\frac{1}{p} + \frac{1}{q} = 1$. Then, for each $g \in L^q(\mu)$, the mapping $\varphi_g : L^p(\mu) \to \mathbb{C}$ defined by

$$\varphi_g(f) := \int_\Omega fg \, d\mu, \quad \text{for } f \in L^p(\mu), \tag{2.10}$$

is a continuous linear functional on $L^p(\mu)$. And, since the mapping $\varphi : L^q(\mu) \to (L^p(\mu))^*$ defined by $\varphi(g) := \varphi_g$, with $\varphi_g(f)$ given by (2.10) for each $f \in L^p(\mu)$, turns out to be a norm isomorphism, [11, pp. 290–292], the dual space of $L^p(\mu)$ can be written explicitly:

$$\left(L^p(\mu)\right)^* = L^q(\mu).$$

Note, for μ being a σ-finite measure, that the dual space of $L^1(\mu)$ can be specified as well. Namely,

$$\left(L^1(\mu)\right)^* = L^\infty(\mu),$$

with each $g \in L^\infty(\mu)$ acting in $L^1(\mu)$ via $f \mapsto \int_\Omega fg \, d\mu$, for $f \in L^1(\mu)$. Moreover, $L^p(\mu)$ is reflexive for $1 < p < \infty$, [35, p. 105].

Let us return to the space of all μ-measurable functions $M(\mu)$ over an arbitrary measure space (Ω, Σ, μ). Concerning convergence in $M(\mu)$ there are two major concepts. A sequence $\{f_n\}_{n \in \mathbb{N}} \subseteq M(\mu)$ is said to *converge μ-a.e.* to $f \in M(\mu)$ if there exists a μ-null set A such that $\{f_n\}_{n \in \mathbb{N}}$ converges pointwise to f on its complement A^c. This definition leads to an equivalent formulation of convergence μ-a.e., [11, p. 250]: As the "diverging set" A is a μ-null set it is clear that, for every $\varepsilon > 0$, the set

$$\left\{w \in \Omega : \forall\, n \in \mathbb{N}\, \exists\, k \in \mathbb{N} : |f_{n+k}(w) - f(w)| \geqslant \varepsilon\right\} \tag{2.11}$$

is μ-null, meaning that

$$\mu\left(\bigcap_{n=1}^{\infty} \bigcup_{k=1}^{\infty} \left\{w \in \Omega : |f_{n+k}(w) - f(w)| \geqslant \varepsilon\right\}\right) = 0, \quad \text{for all } \varepsilon > 0. \tag{2.12}$$

The second important type of convergence of sequences $\{f_n\}_{n \in \mathbb{N}} \subseteq M(\mu)$ which will play a major role in the forthcoming theory is the so-called local convergence in measure. A sequence $\{f_n\}_{n \in \mathbb{N}} \subseteq M(\mu)$ is said to *locally converge in measure* to a function $f \in M(\mu)$ if, for every $\varepsilon > 0$ and for all $A \in \Sigma$ satisfying $\mu(A) < \infty$, we have

$$\lim_{n \to \infty} \mu\left(\left\{w \in \Omega : |f_n(w) - f(w)| \geqslant \varepsilon\right\} \cap A\right) = 0. \tag{2.13}$$

The interesting aspect is that there exists a "topology of local convergence" on $M(\mu)$ generated by a family of pseudo-metrics. For each $A \in \Sigma$ satisfying $\mu(A) < \infty$, define a mapping $\rho_A : M(\mu) \times M(\mu) \to [0, \infty)$ by

$$\rho_A(f, g) := \int_A \frac{|f - g|}{1 + |f - g|} \, d\mu.$$

19

The integral exists since $\mu(A) < \infty$ and $\frac{|f-g|}{1+|f-g|}$ is majorized by χ_Ω. Each of the ρ_A is a pseudo-metric (and not a metric as $\rho_A(f,g) = 0$ may be true for $f \neq g$) and the family

$$\{\rho_A : A \in \Sigma, \ \mu(A) < \infty\}$$

defines a topology on $M(\mu)$, the so-called *topology of local convergence*. Note that the topology can also be generated by an equivalent family of pseudo-metrics, [12, pp. 178–179]. Whenever μ is a σ-finite measure the topology on $M(\mu)$ is metrizable. This may be achieved by a mapping $d : M(\mu) \times M(\mu) \to [0, \infty)$ defined by

$$d(f,g) := \sum_{j=1}^\infty \frac{\rho_{A_j}(f,g)}{2^j(1 + \mu(A_j))} = \sum_{j=1}^\infty \frac{1}{2^j(1 + \mu(A_j))} \int_{A_j} \frac{|f-g|}{1+|f-g|} \, d\mu, \quad (2.14)$$

where $\{A_j\}_{j \in \mathbb{N}} \subseteq \Sigma$ is any sequence of non-null measurable sets satisfying $\bigcup_{j \in \mathbb{N}} A_j = \Omega$ and $\mu(A_j) < \infty$, for all $j \in \mathbb{N}$. Then d is a pseudo-metric, as each of the ρ_{A_j} is a pseudo-metric. Furthermore, $d(f,g) = 0$ if and only if $\rho_{A_j}(f,g) = 0$, for all $j \in \mathbb{N}$, if and only if $f = g$ μ-a.e. on A_j, for all $j \in \mathbb{N}$. But, as $\bigcup_{j \in \mathbb{N}} A_j = \Omega$, this is equivalent to the assertion that $f = g$ μ-a.e. on Ω, i.e., $f = g$ in $M(\mu)$. Thus, d is a metric on $M(\mu)$ and it defines the same topology on $M(\mu)$ as the family of pseudo-metrics $\{\rho_A : A \in \Sigma, \ \mu(A) < \infty\}$.

The next remark gives a description of the convergence in the topology of $M(\mu)$.

Remark 2.2.1

Let (Ω, Σ, μ) be a σ-finite measure space and $\{f_n\}_{n \in \mathbb{N}} \subseteq M(\mu)$ be a sequence of measurable functions. Then the following assertion holds: $\{f_n\}_{n \in \mathbb{N}}$ locally converges in measure to a function $f \in M(\mu)$ if and only if $\lim_{n \to \infty} d(f_n, f) = 0$, with d given by (2.14).

Proof of Remark 2.2.1:

First of all, let us derive some useful inequalities concerning the pseudo-metrics ρ_A. Fix $\varepsilon > 0$ and define, for each $n \in \mathbb{N}$, the set

$$B_{\varepsilon,n} := \{w \in \Omega : |f_n(w) - f(w)| \geqslant \varepsilon\}. \quad (2.15)$$

Here, $\{f_n\}_{n \in \mathbb{N}} \subseteq M(\mu)$ and $f \in M(\mu)$ are arbitrary. Then $B_{\varepsilon,n} \in \Sigma$ and, for each $A \in \Sigma$ satisfying $\mu(A) < \infty$, we obtain $B_{\varepsilon,n} \cap A \in \Sigma$ with $\mu(B_{\varepsilon,n} \cap A) < \infty$. Note that on $B_{\varepsilon,n}$ we have

$$1 \geqslant \frac{|f_n - f|}{1 + |f_n - f|} \geqslant \frac{\varepsilon}{1 + \varepsilon}, \quad (2.16)$$

whereas on $B_{\varepsilon,n}^c$ the inequality becomes

$$\frac{|f_n - f|}{1 + |f_n - f|} < \frac{\varepsilon}{1 + \varepsilon}. \tag{2.17}$$

Hence, for each $n \in \mathbb{N}$, the following inequalities hold, for every $A \in \Sigma$:

$$
\begin{aligned}
\mu(B_{\varepsilon,n} \cap A) &= \int_{B_{\varepsilon,n} \cap A} \chi_\Omega \, d\mu \\
&\overset{(2.16)}{\geqslant} \int_{B_{\varepsilon,n} \cap A} \frac{|f_n - f|}{1 + |f_n - f|} \, d\mu \\
&\overset{(2.16)}{\geqslant} \int_{B_{\varepsilon,n} \cap A} \frac{\varepsilon}{1 + \varepsilon} \, d\mu \\
&= \frac{\varepsilon}{1 + \varepsilon} \mu(B_{\varepsilon,n} \cap A). \tag{2.18}
\end{aligned}
$$

Now, assume that $\{f_n\}_{n \in \mathbb{N}} \subseteq M(\mu)$ is a sequence locally converging in measure to a function $f \in M(\mu)$, meaning that, for each $\varepsilon > 0$ and for each $A \in \Sigma$ satisfying $\mu(A) < \infty$, we have

$$\lim_{n \to \infty} \mu\big(\{w \in \Omega : |f_n(w) - f(w)| \geqslant \varepsilon\} \cap A\big) = 0.$$

Fix any sequence $\{A_j\}_{j \in \mathbb{N}} \subseteq \Sigma$ of measurable sets satisfying $\bigcup_{j \in \mathbb{N}} A_j = \Omega$ and $0 < \mu(A_j) < \infty$, for all $j \in \mathbb{N}$. Then we obtain, for each $\varepsilon > 0$ and each $j \in \mathbb{N}$, that

$$\lim_{n \to \infty} \mu\big(\{w \in \Omega : |f_n(w) - f(w)| \geqslant \varepsilon\} \cap A_j\big) = 0.$$

Fix $\varepsilon > 0$. For each $j \in \mathbb{N}$ this means (see (2.15)) that there exists an index $n_0(\varepsilon, j) \in \mathbb{N}$ such that

$$\mu(B_{\varepsilon,n} \cap A_j) < \varepsilon, \quad \text{for all } n \geqslant n_0(\varepsilon, j). \tag{2.19}$$

Then, for all $n \geqslant n_0(\varepsilon, j)$, we have

$$
\begin{aligned}
\int_{A_j} \frac{|f_n - f|}{1 + |f_n - f|} \, d\mu &= \int_{B_{\varepsilon,n} \cap A_j} \frac{|f_n - f|}{1 + |f_n - f|} \, d\mu + \int_{B_{\varepsilon,n}^c \cap A_j} \frac{|f_n - f|}{1 + |f_n - f|} \, d\mu \\
&\overset{(2.18)}{\leqslant} \mu(B_{\varepsilon,n} \cap A_j) + \int_{B_{\varepsilon,n}^c \cap A_j} \frac{|f_n - f|}{1 + |f_n - f|} \, d\mu \\
&\overset{(2.17)}{<} \mu(B_{\varepsilon,n} \cap A_j) + \int_{B_{\varepsilon,n}^c \cap A_j} \frac{\varepsilon}{1 + \varepsilon} \, d\mu \\
&= \mu(B_{\varepsilon,n} \cap A_j) + \frac{\varepsilon}{1 + \varepsilon} \mu(B_{\varepsilon,n}^c \cap A_j) \\
&\leqslant \mu(B_{\varepsilon,n} \cap A_j) + \frac{\varepsilon}{1 + \varepsilon} \mu(A_j)
\end{aligned}
$$

21

$$\overset{(2.19)}{<} \quad \varepsilon + \varepsilon\,\mu(A_j) \;=\; \varepsilon\,(1+\mu(A_j)).$$

Accordingly,

$$\frac{1}{1+\mu(A_j)} \int_{A_j} \frac{|f_n - f|}{1 + |f_n - f|}\, d\mu \leqslant \varepsilon, \quad \text{for all } n \geqslant n_0(\varepsilon, j).$$

As $\varepsilon > 0$ was chosen arbitrarily, we obtain that

$$\lim_{n \to \infty} \frac{1}{1+\mu(A_j)} \int_{A_j} \frac{|f_n - f|}{1 + |f_n - f|}\, d\mu = 0, \quad \text{for all } j \in \mathbb{N}. \tag{2.20}$$

To show that $\lim_{n \to \infty} d(f_n, f) = 0$ fix an arbitrary $\varepsilon > 0$. Since $\sum_{j=1}^{\infty} \frac{1}{2^j}$ is an absolutely convergent series, we can find an index $j_0 \in \mathbb{N}$ such that

$$\sum_{j=j_0+1}^{\infty} \frac{1}{2^j} < \frac{\varepsilon}{2}.$$

Due to the inequality

$$\frac{1}{2^j(1+\mu(A_j))} \int_{A_j} \frac{|f_n - f|}{1 + |f_n - f|}\, d\mu \;\leqslant\; \frac{1}{2^j(1+\mu(A_j))} \int_{A_j} \chi_\Omega\, d\mu$$

$$= \frac{1}{2^j} \cdot \underbrace{\frac{\mu(A_j)}{1+\mu(A_j)}}_{\leqslant 1} \leqslant \frac{1}{2^j}, \quad \text{for all } n \in \mathbb{N},$$

also

$$\sum_{j=j_0+1}^{\infty} \frac{1}{2^j(1+\mu(A_j))} \int_{A_j} \frac{|f_n - f|}{1 + |f_n - f|}\, d\mu < \frac{\varepsilon}{2} \tag{2.21}$$

is true. On the other hand, (2.20) implies that for the given $\varepsilon > 0$ there exists an index $n_\varepsilon \in \mathbb{N}$ such that

$$\sum_{j=1}^{j_0} \frac{1}{2^j(1+\mu(A_j))} \int_{A_j} \frac{|f_n - f|}{1 + |f_n - f|}\, d\mu < \frac{\varepsilon}{2}, \tag{2.22}$$

for all $n \geqslant n_\varepsilon$. Taking the estimates (2.21) and (2.22) together we obtain that

$$\sum_{j=1}^{\infty} \frac{1}{2^j(1+\mu(A_j))} \int_{A_j} \frac{|f_n - f|}{1 + |f_n - f|}\, d\mu$$

$$= \sum_{j=1}^{j_0} \frac{1}{2^j(1+\mu(A_j))} \int_{A_j} \frac{|f_n - f|}{1 + |f_n - f|}\, d\mu + \sum_{j=j_0+1}^{\infty} \frac{1}{2^j(1+\mu(A_j))} \int_{A_j} \frac{|f_n - f|}{1 + |f_n - f|}\, d\mu$$

$$< \frac{\varepsilon}{2} + \frac{\varepsilon}{2} = \varepsilon, \quad \text{for all } n \geqslant n_\varepsilon.$$

As $\varepsilon > 0$ was chosen arbitrarily we can conclude that

$$\lim_{n\to\infty} d(f_n, f) = \lim_{n\to\infty} \sum_{j=1}^{\infty} \frac{1}{2^j(1 + \mu(A_j))} \int_{A_j} \frac{|f_n - f|}{1 + |f_n - f|} \, d\mu = 0.$$

Conversely, let $\{f_n\}_{n\in\mathbb{N}} \subseteq M(\mu)$ be any sequence of μ-measurable functions satisfying $\lim_{n\to\infty} d(f_n, f) = 0$ for some function $f \in M(\mu)$. Then, by (2.14), we have

$$\lim_{n\to\infty} \frac{1}{1 + \mu(A_j)} \int_{A_j} \frac{|f_n - f|}{1 + |f_n - f|} \, d\mu = 0,$$

and hence,

$$\lim_{n\to\infty} \int_{A_j} \frac{|f_n - f|}{1 + |f_n - f|} \, d\mu = 0,$$

for all $j \in \mathbb{N}$. Here, $\{A_j\}_{j\in\mathbb{N}} \subseteq \Sigma$ is any sequence of measurable sets such that $\bigcup_{j\in\mathbb{N}} A_j = \Omega$ and $\mu(A_j) < \infty$, for all $j \in \mathbb{N}$. Now, let $A \in \Sigma$ be any set satisfying $\mu(A) < \infty$. Choose $\{A_j\}_{j\in\mathbb{N}}$ such that $A = A_{j_0}$ for some $j_0 \in \mathbb{N}$. Define, for $\varepsilon > 0$ fixed, the sets $B_{\varepsilon,n}$ as done in (2.15). Hence, $B_{\varepsilon,n} \cap A \in \Sigma$ and $\mu(B_{\varepsilon,n} \cap A) < \infty$. Considering the inequalities (2.16) we finally obtain (for $A = A_{j_0}$) that

$$0 = \lim_{n\to\infty} \int_{B_{\varepsilon,n}\cap A} \frac{|f_n - f|}{1 + |f_n - f|} \, d\mu \geqslant \lim_{n\to\infty} \frac{\varepsilon}{1 + \varepsilon} \mu(B_{\varepsilon,n} \cap A) \geqslant 0,$$

meaning that $\{f_n\}_{n\in\mathbb{N}}$ locally converges in measure to f. $\quad\square$

The following remark states how convergence μ-a.e. and local convergence in measure are linked together as soon as (Ω, Σ, μ) is a σ-finite measure space.

Remark 2.2.2

Let (Ω, Σ, μ) be a σ-finite measure space. Then the following assertions hold:

(i) Let $\{f_n\}_{n\in\mathbb{N}} \subseteq M(\mu)$ be a sequence of μ-measurable functions locally converging in measure to $f \in M(\mu)$ as well as to $g \in M(\mu)$. Then $f = g$ locally μ-a.e., meaning that, for each $A \in \Sigma$ satisfying $\mu(A) < \infty$, the equality $f\chi_A = g\chi_A$ holds μ-a.e., [11, p. 254]. Since the measure space is σ-finite it follows that $f = g$ μ-a.e..

(ii) Let $\{f_n\}_{n\in\mathbb{N}} \subseteq M(\mu)$ be a sequence of μ-measurable functions converging μ-a.e. to $f \in M(\mu)$. Then $\{f_n\}_{n\in\mathbb{N}}$ locally converges in measure to f, [12, p. 174].

(iii) A sequence $\{f_n\}_{n\in\mathbb{N}} \subseteq M(\mu)$ of μ-measurable functions locally converges in measure to f if and only if each subsequence of $\{f_n\}_{n\in\mathbb{N}}$ admits a subsequence which converges μ-a.e. to f, [11, p. 258].

(iv) $M(\mu)$ is complete for the topology of local convergence, [33, p. 268–269]. □

Note that the Σ-simple functions $\mathrm{sim}(\Sigma)$ are dense in $M(\mu)$, [11, p. 242].

Further important theorems concerning convergence of a sequence of μ-measurable functions are the following, [11, p. 125 & p. 145].

Proposition 2.2.1 (Monotone Convergence Theorem)
Let $\{f_n\}_{n\in\mathbb{N}} \subseteq M(\mu)^+$ be any increasing sequence of functions. Then,

$$\int_\Omega \left(\lim_{n\to\infty} f_n\right) d\mu = \lim_{n\to\infty} \int_\Omega f_n \, d\mu. \quad \square$$

Proposition 2.2.2 (Lebesgue's Dominated Convergence Theorem)
Let $\{f_n\}_{n\in\mathbb{N}} \subseteq M(\mu)$ be a sequence of functions converging μ-a.e. to a function $f \in M(\mu)$. Whenever there exists a μ-integrable function $g \in M(\mu)^+$ satisfying $|f_n| \leqslant g$ μ-a.e. on Ω, for all $n \in \mathbb{N}$, then also the functions f, f_n are μ-integrable, for all $n \in \mathbb{N}$, and

$$\lim_{n\to\infty} \int_\Omega f_n \, d\mu = \int_\Omega f \, d\mu. \quad \square$$

Another important theorem in measure theory is Fubini's Theorem. For the theory of integration with respect to product measures see, for example, [11, pp. 164–191].

Proposition 2.2.3 (Fubini's Theorem)
Let (X, Σ_X, μ), (Y, Σ_Y, ν) be σ-finite measure spaces and denote by $X \times Y$ the product space of the topological spaces X and Y, by $\Sigma_X \otimes \Sigma_Y$ the product σ-algebra and by $\mu \otimes \nu$ the product measure of μ and ν. Then the following assertions hold.

(i) For each non-negative $\Sigma_X \otimes \Sigma_Y$-measurable function $f : X \times Y \to [0, \infty)$ the functions defined on X resp. Y by

$$x \mapsto \int_Y f(x, y) \, d\nu(y) \quad \text{and} \quad y \mapsto \int_X f(x, y) \, d\mu(x)$$

are Σ_X-measurable resp. Σ_Y-measurable and

$$\int_{X\times Y} f \, d\mu \otimes \nu = \int_X \left(\int_Y f(x, y) \, d\nu(y) \right) d\mu(x)$$

24

$$= \int_Y \left(\int_X f(x,y)\, d\mu(x) \right) d\nu(y).$$

(ii) Let $f : X \times Y \to \mathbb{C}$ be $\mu \otimes \nu$-integrable. Then $f(x, \cdot)$ is ν-integrable for μ-almost every $x \in X$ and

$$A := \{ x \in X : f(x, \cdot) \text{ is not } \nu\text{-integrable} \} \in \Sigma_X;$$

respectively, $f(\cdot, y)$ is μ-integrable for ν-almost every $y \in Y$ and

$$B := \{ y \in Y : f(\cdot, y) \text{ is not } \mu\text{-integrable} \} \in \Sigma_Y.$$

Moreover, the functions

$$x \mapsto \int_Y f(x,y)\, d\nu(y) \quad \text{and} \quad y \mapsto \int_X f(x,y)\, d\mu(x)$$

are μ-integrable over A^c resp. ν-integrable over B^c, and

$$\int_{X \times Y} f\, d\mu \otimes \nu = \int_{A^c} \left(\int_Y f(x,y)\, d\nu(y) \right) d\mu(x)$$

$$= \int_{B^c} \left(\int_X f(x,y)\, d\mu(x) \right) d\nu(y). \quad \square$$

A set function $\mu : \Sigma \to \mathbb{C}$ is called a *complex measure* if $\mu(\varnothing) = 0$ and if it is σ-additive, that is,

$$\mu \left(\bigcup_{j=1}^{\infty} A_j \right) = \sum_{j=1}^{\infty} \mu(A_j),$$

for any sequence $\{A_j\}_{j \in \mathbb{N}} \subseteq \Sigma$ of disjoint sets. Associated with the complex measure μ define a set function $|\mu| : \Sigma \to [0, \infty)$ by

$$|\mu|(A) := \sup_{\pi} \sum_{j=1}^{l} |\mu(A_j)|$$

where the supremum is taken over all finite partitions $\pi = \{A_j\}_{j=1}^{l}$ of $A \in \Sigma$. Then $|\mu|$ is called the *variation measure* of μ and the finite number $\|\mu\| := |\mu|(\Omega)$ is called the *total variation* of μ. Concerning the variation of a measure, the following proposition turns out to be a useful tool, [31, p. 152].

Proposition 2.2.4

Let μ be a positive measure on Σ, $g \in L^1(\mu)$ and define $\lambda(A) = \int_A g\, d\mu$, for all $A \in \Sigma$.

25

Then the following equality holds:

$$|\lambda|(A) = \int_A |g|\, d\mu, \quad \text{for } A \in \Sigma. \quad \square$$

Moreover, some further properties of the variation of a complex measure are brought together in the following lemma; see [31, Chapter 6] or [10, Chapter III], for instance.

Lemma 2.2.1

Let $\mu : \Sigma \to \mathbb{C}$ be a complex measure. Then the following assertions hold:

(i) $|\mu(A)| \leqslant |\mu|(A)$, for all $A \in \Sigma$.

(ii) $|\mu|(B) \leqslant |\mu|(A)$, for all $A, B \in \Sigma$ with $B \subseteq A$.

(iii) $\sup\{|\mu(B)| : B \in \Sigma, B \subseteq A\} \leqslant |\mu|(A) \leqslant 4 \sup\{|\mu(B)| : B \in \Sigma, B \subseteq A\}$,
for all $A \in \Sigma$. $\quad \square$

Finally, a measure $\mu : \Sigma \to [0, \infty]$ is said to be *non-atomic* if, for each $A \in \Sigma$ satisfying $\mu(A) > 0$, there exists a set $B \in A \cap \Sigma := \{A \cap S : S \in \Sigma\}$ such that $\mu(B) \neq 0$ and $\mu(A \backslash B) \neq 0$. The range of a finite, positive non-atomic measure $\mu : \Sigma \to [0, \infty)$ is the closed interval $[0, \mu(\Omega)]$; see [16] and the references therein. Moreover, we say that μ has the *Darboux property* on Σ if, for each $A \in \Sigma$ and $0 < t < \mu(A)$, there exists a set $B \in A \cap \Sigma$ such that $\mu(B) = t$. Note that a non-atomic measure always has the Darboux property, [16]. Non-atomic measures are in some ways advantageous as the following lemma shows.

Lemma 2.2.2

Let $\mu : \Sigma \to [0, \infty)$ be a finite, positive non-atomic measure. Let $A \in \Sigma$ be such that $\mu(A) > 0$ and let $l \in \mathbb{N}$ be fixed. Then there exists a partition $\{A_j\}_{j=1}^{l} \subseteq \Sigma$ of A such that

$$\mu(A_j) = \frac{\mu(A)}{l}, \quad \text{for all } j = 1, \ldots, l.$$

Proof:

Choose an arbitrary $A \in \Sigma$ satisfying $\mu(A) > 0$ and fix $l \in \mathbb{N}$. Let $\mu(A) =: \alpha$. Since μ is a finite, positive non-atomic measure, μ has the Darboux property on Σ. Thus, there exists a set $A_1 \in A \cap \Sigma \subseteq \Sigma$ such that $\mu(A_1) = \frac{\alpha}{l}$. The additivity of μ gives

$$\alpha = \mu(A) = \mu\big(A_1 \cup \underbrace{(A \backslash A_1)}_{=:B_1}\big) = \mu(A_1) + \mu(B_1) = \tfrac{\alpha}{l} + \mu(B_1)$$

26

or, equivalently,

$$\mu(B_1) = \alpha - \tfrac{\alpha}{l} = (l-1) \cdot \tfrac{\alpha}{l}.$$

Since $B_1 \in A \cap \Sigma$ and μ restricted to $A \cap \Sigma$ still has the Darboux property we can find a set $A_2 \in B_1 \cap \Sigma \subseteq \Sigma$ such that $\mu(A_2) = \tfrac{\alpha}{l}$. Thereby we obtain that

$$\alpha = \mu(A) = \mu\big(A_1 \cup A_2 \cup \underbrace{(B_1 \backslash A_2)}_{=:B_2}\big) = \mu(A_1) + \mu(A_2) + \mu(B_2) = 2 \cdot \tfrac{\alpha}{l} + \mu(B_2)$$

or, equivalently,

$$\mu(B_2) = \alpha - 2 \cdot \tfrac{\alpha}{l} = (l-2) \cdot \tfrac{\alpha}{l}.$$

Continue inductively and suppose that we have already found sets $A_1, \ldots, A_{l-1} \in \Sigma$ satisfying $A_j \in B_{j-1} \cap \Sigma \subseteq \Sigma$ where $B_0 := A$ and $B_j := B_{j-1} \backslash A_j$, for all $j = 1, \ldots, l-1$, and $\mu(A_j) = \tfrac{\alpha}{l}$, for all $j = 1, \ldots, l-1$. Then,

$$\alpha = \mu(A) = \mu\bigg(\bigcup_{j=1}^{l-1} A_j \cup \underbrace{\big(B_{l-2} \backslash A_{l-1}\big)}_{=:B_{l-1}}\bigg) = \sum_{j=1}^{l-1} \mu(A_j) + \mu(B_{l-1}) = (l-1) \cdot \tfrac{\alpha}{l} + \mu(B_{l-1})$$

or, equivalently,

$$\mu(B_{l-1}) = \alpha - (l-1) \cdot \tfrac{\alpha}{l} = \tfrac{\alpha}{l}.$$

Let $A_l := B_{l-1}$. Then $A_l \in B_{l-1} \cap \Sigma \subseteq \Sigma$ and $\mu(A_l) = \tfrac{\alpha}{l}$ and we have found a partition $\{A_j\}_{j=1}^l \subseteq \Sigma$ of A satisfying $\mu(A_j) = \tfrac{\alpha}{l}$, for all $j = 1, \ldots, l$. □

2.3 Fréchet function spaces

A vector space X over the scalar field $\mathbb{R}$ is called a *Riesz space* or *vector lattice* if it is endowed with a partial order $\leqslant$ such that, for any $x, y \in X$ and $\lambda \in \mathbb{R}$, the following conditions are satisfied:

(i)   If $x \leqslant y$, then $x + z \leqslant y + z$, for all $z \in X$.

(ii)   If $0 \leqslant \lambda$ and $x \leqslant y$, then $\lambda x \leqslant \lambda y$.

(iii)   For any pair of vectors $x, y \in X$ there exists a supremum (denoted by $x \vee y$) in X with respect to the partial order of the lattice structure $\leqslant$.

The element $|x| := x \vee (-x)$ is called the *modulus* of x. The set $X^+ := \{x \in X : 0 \leqslant x\}$ is called the *positive cone* of X. Since $x \leqslant x \vee (-x)$ it follows from (i) that $0 = -x + x \leqslant -x + |x|$. Similarly, $-x \leqslant x \vee (-x)$ implies $0 \leqslant x + |x|$. But, from (i) if $w \leqslant y$ and $u \leqslant v$, then $w + u \leqslant y + v$. It follows that $0 + 0 \leqslant (-x + |x|) + (x + |x|) = 2|x|$. Then (ii) yields $0 \leqslant |x|$, which is valid for every $x \in X$.

A vector lattice is said to be *Archimedean* if $x, y \in X$ and $nx \leqslant y$, for all $n \in \mathbb{N}$,

27

imply that $x \leqslant 0$. A vector subspace I of a Riesz space X is called an *ideal* if it is *solid*, meaning that if $x \in I$ and $y \in X$ satisfy $|y| \leqslant |x|$, then $y \in I$. A *locally solid Riesz space* is a Riesz space X equipped with a locally solid topology τ, meaning that τ has a neighbourhood base at zero consisting of solid sets.

Let (Ω, Σ, μ) be a measure space. A mapping q defined on $M(\mu)^+$ is called a *function semi-norm* whenever it satisfies the following conditions:

(i) $0 \leqslant q \leqslant \infty$.

(ii) If $u = 0$ μ-a.e., then $q(u) = 0$.

(iii) $q(\lambda u) = \lambda \, q(u)$, for every constant $0 \leqslant \lambda < \infty$.

(iv) $q(u + v) \leqslant q(u) + q(v)$, for all $u, v \in M(\mu)^+$.

(v)   If $u, v \in M(\mu)^+$ and $u \leqslant v$, then $q(u) \leqslant q(v)$.

By setting $q(f) := q(|f|)$, a function semi-norm can be extended to the whole of $M(\mu)$. Recall that $M(\mu)$ consists of $\mathbb{C}$-valued μ-measurable functions. In particular, (iii) then implies that

$$q(\lambda f) = q(|\lambda f|) = q(|\lambda||f|) = |\lambda| \, q(|f|) = |\lambda| \, q(f),$$

for every $\lambda \in \mathbb{C}$ and $f \in M(\mu)$.

In the sequel, we will consider a sequence $\{q_k\}_{k \in \mathbb{N}}$ of function semi-norms instead of a single function semi-norm q. A sequence of function semi-norms $\{q_k\}_{k \in \mathbb{N}}$ is called *fundamental* if, whenever $f \in M(\mu) \backslash \{0\}$, there exists an index $m \in \mathbb{N}$ such that $q_m(f) \neq 0$, i.e., $q_m(f) \in (0, \infty]$. We assume $\{q_k\}_{k \in \mathbb{N}}$ to be increasing and fundamental. Define

$$L_{\{q_k\}} := \big\{ f \in M(\mu) : q_k(f) < \infty, \text{ for all } k \in \mathbb{N} \big\} = \bigcap_{k \in \mathbb{N}} L_{q_k},$$

where $L_{q_k} := \{ f \in M(\mu) : q_k(f) < \infty \}$, for $k \in \mathbb{N}$. Then $L_{\{q_k\}}$ is a locally solid, metrizable, locally convex Hausdorff space for the topology induced by $\{q_k\}_{k \in \mathbb{N}}$. Hence, whenever $f \in M(\mu)$ and $g \in L_{\{q_k\}}$ satisfy $|f| \leqslant |g|$, then $f \in L_{\{q_k\}}$ and $q_k(f) \leqslant q_k(g)$, for all $k \in \mathbb{N}$. The space $L_{\{q_k\}}$ is called a (locally solid) *metrizable function space* and, if it is complete, a *Fréchet function space*. The *positive cone* of a Fréchet function space $L_{\{q_k\}}$ is defined by

$$L_{\{q_k\}}^+ := \big\{ f \in L_{\{q_k\}} : f \geqslant 0 \big\}$$

consisting of all those functions in $L_{\{q_k\}}$ that are $[0, \infty)$-valued μ-a.e. on Ω.

Let $\{q_k\}_{k\in\mathbb{N}}$ be an increasing, fundamental sequence of function semi-norms and let $L_{\{q_k\}}$ be defined as above. The metrizable function space $L_{\{q_k\}}$ is said to have the *joint Riesz-Fischer property* (briefly: (JRF)-property) if, given any sequence $\{f_n\}_{n\in\mathbb{N}} \subseteq L_{\{q_k\}}$ satisfying

$$\sum_{n=1}^{\infty} q_k(f_n) < \infty, \text{ for all } k \in \mathbb{N}, \text{ it follows that } q_k\left(\sum_{n=1}^{\infty} |f_n|\right) < \infty, \text{ for all } k \in \mathbb{N},$$

(2.23)

[6, Definition 3.5]. In the definition of the (JRF)-property it is not assumed that $L_{\{q_k\}}$ is complete.

Remark 2.3.1

The above definition of the (JRF)-property can be replaced by an equivalent formulation: $L_{\{q_k\}}$ has the (JRF)-property if, given any sequence $\{u_n\}_{n\in\mathbb{N}} \subseteq L^+_{\{q_k\}}$ satisfying

$$\sum_{n=1}^{\infty} q_k(u_n) < \infty, \text{ for all } k \in \mathbb{N}, \text{ it follows that } q_k\left(\sum_{n=1}^{\infty} u_n\right) < \infty, \text{ for all } k \in \mathbb{N}.$$

(2.24)

To prove Remark 2.3.1 we first need to consider another collection of μ-measurable functions. Denote by $\tilde{M}(\mu)$ the set of all measurable functions $f : \Omega \to \mathbb{R} \cup \{-\infty, +\infty\}$. (For the properties of $\tilde{M}(\mu)$ see, for instance, [11, pp. 104–108].) As usual $\tilde{M}(\mu)^+$ will denote the non-negative functions in $\tilde{M}(\mu)$, with ∞ allowed as a possible value. Now we can draw the following conclusion.

Lemma 2.3.1

Let $\{q_k\}_{k\in\mathbb{N}}$ be an increasing sequence of function semi-norms in $\tilde{M}(\mu)$ (which is fundamental in $M(\mu)$) and let $f \in \tilde{M}(\mu)^+$ satisfy $q_k(f) < \infty$, for all $k \in \mathbb{N}$. Then f is $[0,\infty)$-valued μ-a.e..

Proof:

Let $f \in \tilde{M}(\mu)^+$ satisfy $q_k(f) < \infty$, for all $k \in \mathbb{N}$, and define $A_f := \{w \in \Omega : f(w) = \infty\}$. Then, for each $w \in \Omega$, the inequality $f(w) \geqslant n\chi_{A_f}(w)$ holds, for all $n \in \mathbb{N}$. Each q_k being a function semi-norm we obtain

$$0 \leqslant q_k(n\chi_{A_f}) = n\,q_k(\chi_{A_f}) \leqslant q_k(f) < \infty, \quad \text{for all } k \in \mathbb{N}, \text{ for all } n \in \mathbb{N}.$$

But, since $\mathbb{R}$ is Archimedean, this shows that $q_k(\chi_{A_f}) = 0$, for all $k \in \mathbb{N}$. Hence, we can conclude that $\chi_{A_f} = 0$ μ-a.e.. Therefore, A_f is a μ-null set and $f(w) < \infty$, for

29

μ-almost every $w \in \Omega$. □

Proof of Remark 2.3.1:

It is clear that (2.23) implies (2.24).

Let now (2.24) be valid and let $\{f_n\}_{n\in\mathbb{N}} \subseteq L_{\{q_k\}}$ be any sequence satisfying $\sum_{n=1}^{\infty} q_k(f_n) < \infty$, for all $k \in \mathbb{N}$. Fix an arbitrary $n \in \mathbb{N}$. Since f_n is $\mathbb{C}$-valued, we can write $f_n = g_n + i\, h_n$, where g_n is the real part and h_n the imaginary part of f_n. Since $|g_n| \leqslant |f_n|$ holds and each q_k is a function semi-norm, we obtain

$$q_k(g_n) \leqslant q_k(f_n), \quad \text{for all } k \in \mathbb{N}.$$

On the other hand, g_n is an $\mathbb{R}$-valued function, and so we can write $g_n = g_n^+ - g_n^-$. Hence, we have $|g_n^+| \leqslant |g_n|$ and therefore

$$q_k(g_n^+) \leqslant q_k(g_n) \leqslant q_k(f_n), \quad \text{for all } k \in \mathbb{N}.$$

As n was chosen arbitrarily, this is true for all $n \in \mathbb{N}$, meaning that $q_k(f_n)$ is a majorant of $q_k(g_n^+)$. Thus,

$$\sum_{n=1}^{\infty} q_k(g_n^+) \leqslant \sum_{n=1}^{\infty} q_k(f_n) < \infty, \quad \text{for all } k \in \mathbb{N}.$$

The condition (2.24) implies that

$$q_k\left(\sum_{n=1}^{\infty} g_n^+\right) < \infty, \quad \text{for all } k \in \mathbb{N},$$

and Lemma 2.3.1 yields that $\sum_{n=1}^{\infty} g_n^+ < \infty$ μ-a.e. on Ω. By repeating the arguments we obtain the corresponding results for g_n^-, h_n^+ and h_n^-. But, as

$$|f_n| = |g_n + i\, h_n| \leqslant |g_n| + |h_n| = g_n^+ + g_n^- + h_n^+ + h_n^-,$$

for all $n \in \mathbb{N}$, it follows that

$$\sum_{n=1}^{\infty} |f_n| \leqslant \sum_{n=1}^{\infty} g_n^+ + \sum_{n=1}^{\infty} g_n^- + \sum_{n=1}^{\infty} h_n^+ + \sum_{n=1}^{\infty} h_n^-,$$

by which we can finally conclude, by the triangle inequality for each q_k, that

$$q_k\left(\sum_{n=1}^{\infty} |f_n|\right) \leqslant q_k\left(\sum_{n=1}^{\infty} g_n^+\right) + q_k\left(\sum_{n=1}^{\infty} g_n^-\right) + q_k\left(\sum_{n=1}^{\infty} h_n^+\right) + q_k\left(\sum_{n=1}^{\infty} h_n^-\right) < \infty,$$

for all $k \in \mathbb{N}$. Hence, also (2.23) holds, that is, $L_{\{q_k\}}$ has the (JRF)-property. $\square$

In the following remark and lemma, we will denote Fréchet function spaces $L_{\{q_k\}}$ over a measure space (Ω, Σ, μ) by $X(\mu)$ and their positive cone by $X(\mu)^+$. An element $f \in X(\mu)$ is said to be *σ-order continuous* (briefly, σ-o.c.) if it has the property that a sequence $\{u_n\}_{n \in \mathbb{N}} \subseteq X(\mu)^+$ converges to 0 in the topology of $X(\mu)$ whenever it satisfies $|f| \geqslant u_n \downarrow_n 0$ pointwise μ-a.e. on Ω. The collection of all σ-o.c. elements of $X(\mu)$ is called the *σ-order continuous part* of $X(\mu)$ and is denoted by $X(\mu)_a$.

The next remark occurs in [6, Lemma 3.11]. Since the manuscript is unpublished we give the proof of the statement here again.

Remark 2.3.2

Let $f \in X(\mu)$. Then $f \in X(\mu)_a$ if and only if for every sequence $\{f_n\}_{n \in \mathbb{N}} \subseteq X(\mu)$ with $|f_n| \leqslant |f|$ and for which $\lim_{n \to \infty} f_n = f_0$ exists pointwise μ-a.e., it follows that $\{f_n\}_{n \in \mathbb{N}}$ converges to f_0 in the topology of $X(\mu)$.

Proof of Remark 2.3.2:

Let $f \in X(\mu)_a$. Choose any sequence $\{f_n\}_{n \in \mathbb{N}}$ in $X(\mu)$ with $|f_n| \leqslant |f|$ for which $\lim_{n \to \infty} f_n = f_0$ pointwise. Since $|f_0| \leqslant |f|$, we have f_0 in $X(\mu)$. Then the sequence $\{u_n\}_{n \in \mathbb{N}}$ defined by $u_n := \sup\{|f_j - f_0| : j \geqslant n\}$, for all $n \in \mathbb{N}$, satisfies both $u_n \leqslant 2|f|$, for all $n \in \mathbb{N}$ (in particular, $\{u_n\}_{n \in \mathbb{N}} \subseteq X(\mu)^+$), and $u_n \downarrow_n 0$. Hence, $\{u_n\}_{n \in \mathbb{N}}$ converges to 0 in the topology of $X(\mu)$. Since $|f_n - f_0| \leqslant u_n$, for all $n \in \mathbb{N}$, and the topology of $X(\mu)$ is locally solid, it follows that $\{f_n\}_{n \in \mathbb{N}}$ converges to f_0 in the topology of $X(\mu)$.

The converse statement is obvious. $\square$

A Fréchet function space $X(\mu)$ is said to have a *σ-Lebesgue topology* if it has the property that a sequence $\{u_n\}_{n \in \mathbb{N}} \subseteq X(\mu)^+$ converges to 0 in the topology of $X(\mu)$ whenever it satisfies $u_n \downarrow_n 0$ pointwise μ-a.e. on Ω. It is clear that, for every Fréchet function space having a σ-Lebesgue topology, the σ-order continuous part and the space itself coincide, i.e., $X(\mu) = X(\mu)_a$.

The next result emphasizes the importance of the σ-Lebesgue topology for the theory in the forthcoming chapters.

Lemma 2.3.2

Let $X(\mu)$ be a Fréchet function space containing the Σ-simple functions $\mathrm{sim}(\Sigma)$ and having a σ-Lebesgue topology. Given $f \in X(\mu)$, there exists a sequence $\{r_n\}_{n \in \mathbb{N}} \subseteq$

$sim(\Sigma)$ such that $|r_n| \leqslant |f|$, for $n \in \mathbb{N}$, with $\{r_n\}_{n\in\mathbb{N}}$ converging to f both pointwise on Ω and in the topology of $X(\mu)$. In particular, $sim(\Sigma)$ is dense in $X(\mu)$.

Proof:

Let $u \in X(\mu)^+$. Choose a sequence of functions $\{s_n\}_{n\in\mathbb{N}} \subseteq sim(\Sigma)$ satisfying $0 \leqslant s_n \uparrow_n u$ pointwise on Ω. Then $(u - s_n) \downarrow_n 0$. As $X(\mu)$ has a σ-Lebesgue topology, it follows that

$$\lim_{n\to\infty} q_k(u - s_n) = 0, \quad \text{for all } k \in \mathbb{N},$$

meaning that $\{s_n\}_{n\in\mathbb{N}}$ converges to u in the topology of $X(\mu)$. For an arbitrary $f \in X(\mu)$, note that $f = (g^+ - g^-) + i\,(h^+ - h^-)$ where $g = g^+ - g^-$ denotes the real part and $h = h^+ - h^-$ the imaginary part of f. As $g^+, g^-, h^+, h^- \in X(\mu)^+$, there exist sequences $\{s_n\}_{n\in\mathbb{N}}$, $\{t_n\}_{n\in\mathbb{N}}$ of $\mathbb{R}$-valued Σ-simple functions such that $0 \leqslant s_n^+ \uparrow g^+, 0 \leqslant s_n^- \uparrow g^-$ and $0 \leqslant t_n^+ \uparrow h^+, 0 \leqslant t_n^- \uparrow h^-$ pointwise on Ω as well as in the topology of $X(\mu)$. Define

$$r_n := (s_n^+ - s_n^-) + i\,(t_n^+ - t_n^-), \quad \text{for } n \in \mathbb{N},$$

then $\{r_n\}_{n\in\mathbb{N}} \subseteq sim(\Sigma)$ with $|r_n| \leqslant |f|$, for $n \in \mathbb{N}$, and, by the triangle inequality, we have

$$|f - r_n| \leqslant |g^+ - s_n^+| + |g^- - s_n^-| + |h^+ - t_n^+| + |h^- - t_n^-|,$$

for all $n \in \mathbb{N}$. Each q_k being a function semi-norm we finally obtain, by the triangle inequality for each q_k, that

$$q_k(f - r_n) \leqslant q_k(g^+ - s_n^+) + q_k(g^- - s_n^-) + q_k(h^+ - t_n^+) + q_k(h^- - t_n^-),$$

for all $k \in \mathbb{N}$. Taking the limit on both sides we derive

$$\lim_{n\to\infty} q_k(f - r_n) = 0, \quad \text{for all } k \in \mathbb{N}.$$

Hence, there exists a sequence of Σ-simple functions $\{r_n\}_{n\in\mathbb{N}} \subseteq sim(\Sigma)$ with $|r_n| \leqslant |f|$, for $n \in \mathbb{N}$, converging pointwise to f and in the topology of $X(\mu)$. $\square$

Let us give two examples of Fréchet function spaces that we will extensively make use of in Chapter 4.

Example 2.3.1

The Fréchet function space $L^{p-}([0,1])$

Let $\Omega := [0,1]$ and let λ be Lebesgue measure. In that case $([0,1], \mathcal{B}([0,1]), \lambda)$ is a finite measure space. Denote by $L^0([0,1])$ the Lebesgue measurable functions $f :$ $[0,1] \to \mathbb{C}$. Fix $p \in (1,\infty)$ and let $\{r_k\}_{k \in \mathbb{N}}$ be any sequence of real numbers in $[1,p)$ satisfying $1 \leqslant r_k \uparrow_k p$. Define, for each $k \in \mathbb{N}$, a mapping $q_k : L^0([0,1]) \to [0,\infty]$ by

$$q_k(f) := \left(\int_0^1 |f|^{r_k} \, d\lambda \right)^{1/r_k}.$$

Then each q_k is a function semi-norm, even a function norm on $L^0([0,1])$. (Observe that L_{q_k} is the usual Banach space $L^{r_k}([0,1])$, for $k \in \mathbb{N}$.) This follows from [11, p. 224] and the monotonicity of the Lebesgue integral, [11, p. 132]: Let $f, g \in L^0([0,1])$ satisfy $|f| \leqslant |g|$, then $|f|^{r_k} \leqslant |g|^{r_k}$ and consequently

$$q_k(f) = \left(\int_0^1 |f|^{r_k} \, d\lambda \right)^{1/r_k} \leqslant \left(\int_0^1 |g|^{r_k} \, d\lambda \right)^{1/r_k} = q_k(g),$$

for all $k \in \mathbb{N}$. On the other hand, Hölder's inequality (2.9) implies that the sequence of function norms $\{q_k\}_{k \in \mathbb{N}}$ is increasing. To see this, fix $k \in \mathbb{N}$ and let $f \in L^0([0,1])$. Then $|f|^{r_k} \in L^0([0,1])$ and by setting $r := \frac{r_{k+1}}{r_k}$ and $s := \left(1 - \frac{1}{r}\right)^{-1}$ we obtain $\frac{1}{r} + \frac{1}{s} = 1$. Applying Hölder's inequality gives

$$\int_0^1 |f|^{r_k} \, d\lambda \overset{(2.9)}{\leqslant} \left(\int_0^1 |f|^{r_k r} \, d\lambda \right)^{1/r} \left(\int_0^1 |\chi_{[0,1]}|^s d\lambda \right)^{1/s}$$

$$= \left(\int_0^1 |f|^{r_{k+1}} d\lambda \right)^{r_k/r_{k+1}} \lambda([0,1])^{1/s}$$

meaning that $q_k(f) \leqslant q_{k+1}(f)$; see also (2.8). Moreover, it is clear that for each $0 \neq f \in L^0([0,1])$ there exists an index $m \in \mathbb{N}$ such that $q_m(f) \in (0,\infty]$. Thus, we obtain an increasing fundamental sequence of function norms in the metrizable function space $L^{p-}([0,1])$ defined by

$$L^{p-}([0,1]) = \bigcap_{k \in \mathbb{N}} \{f \in L^0([0,1]) : q_k(f) < \infty\} = \bigcap_{k \in \mathbb{N}} L^{r_k}([0,1]).$$

Note that $L^{p-}([0,1])$ is also complete. Indeed, for each $k \in \mathbb{N}$, the space L_{q_k} is the usual Banach space $L^{r_k}([0,1])$, and so we have a system of Banach spaces satisfying

$$L^{r_1}([0,1]) \supseteq L^{r_2}([0,1]) \supseteq \ldots \supseteq L^{p-}([0,1]).$$

Because $L^{p-}([0,1])$ coincides with the countable intersection of the Banach spaces $L^{r_k}([0,1])$, where $k \in \mathbb{N}$, we can conclude by [13, pp. 17–18] that $L^{p-}([0,1])$ is complete. Hence, $L^{p-}([0,1])$ becomes a Fréchet function space whose topology is generated by $\{q_k\}_{k \in \mathbb{N}}$. Let us state some properties of the space $L^{p-}([0,1])$:

(i) $\left(L^{p-}([0,1])\right)^* = \bigcup_{k\in\mathbb{N}} L^{s_k}([0,1])$, where $\frac{1}{r_k} + \frac{1}{s_k} = 1$, for $k \in \mathbb{N}$.

Proof:

Since each $L^{r_k}([0,1])$ is a Banach space, $L^{p-}([0,1])$ is by definition a so-called countably normed space. According to [13, p. 36] the dual space of $L^{p-}([0,1])$ can be specified as

$$\left(L^{p-}([0,1])\right)^* = \left(\bigcap_{k\in\mathbb{N}} L^{r_k}([0,1])\right)^* = \bigcup_{k\in\mathbb{N}} \left(L^{r_k}([0,1])\right)^* = \bigcup_{k\in\mathbb{N}} L^{s_k}([0,1]). \quad \square$$

Note that the duality of $L^{p-}([0,1])$ and its dual space is expressed by the bilinear form

$$\langle f, g\rangle := \int_0^1 fg \, d\lambda,$$

for $f \in L^{p-}([0,1])$, $g \in \left(L^{p-}([0,1])\right)^*$.

(ii) $L^{p-}([0,1])$ is reflexive.

Proof:

Since each of the local Banach spaces $L^{r_k}([0,1]) = L^{p-}([0,1])/q_k^{-1}(\{0\})$, for $1 < r_k < p$, is reflexive, Proposition 2.1.2 implies that also $L^{p-}([0,1])$ is reflexive. $\quad \square$

(iii) $L^\infty([0,1]) \subseteq L^{p-}([0,1])$. In particular, $\text{sim}\left(\mathcal{B}([0,1])\right) \subseteq L^{p-}([0,1])$.

Proof:

Let $f \in L^\infty([0,1])$. Then $|f(w)| < M$, for λ-almost every $w \in [0,1]$, for some real number $M > 0$. Thereby, for each $k \in \mathbb{N}$, we obtain

$$q_k(f) = \left(\int_0^1 |f(w)|^{r_k} \, d\lambda(w)\right)^{1/r_k} \leqslant \left(\int_0^1 M^{r_k} d\lambda(w)\right)^{1/r_k}$$
$$= M \, \lambda([0,1])^{1/r_k} < \infty.$$

Hence, $L^\infty([0,1]) \subseteq L^{p-}([0,1])$. Since each $\mathcal{B}([0,1])$-simple function is bounded on $[0,1]$, it follows that $\text{sim}\left(\mathcal{B}([0,1])\right) \subseteq L^\infty([0,1])$ and consequently, also $\text{sim}\left(\mathcal{B}([0,1])\right) \subseteq L^{p-}([0,1])$. $\quad \square$

(iv) $L^{p-}([0,1])$ has a σ-Lebesgue topology.

Proof:

This follows from Lebesgue's Dominated Convergence Theorem 2.2.2. Let $\{u_n\}_{n\in\mathbb{N}} \subseteq L^{p-}([0,1])^+$ be a sequence of functions satisfying $u_n \downarrow_n 0$ pointwise on $[0,1]$. Then we have $0 \leqslant u_n \leqslant u_1$, for all $n \in \mathbb{N}$. Fix $k \in \mathbb{N}$. Then also

34

$0 \leqslant u_n^{r_k} \leqslant u_1^{r_k}$ with $u_n^{r_k} \downarrow_n 0$ pointwise. Since $u_1 \in L^{r_k}([0,1])$, we have that $u_1^{r_k} \in L^1([0,1])$. Thus,

$$\lim_{n\to\infty} q_k^{r_k}(u_n) = \lim_{n\to\infty} \int_0^1 |u_n|^{r_k} \, d\lambda = \int_0^1 \left(\lim_{n\to\infty} |u_n|^{r_k} \right) d\lambda = 0,$$

and hence, also $\lim_{n\to\infty} q_k(u_n) = 0$. Since $k \in \mathbb{N}$ is arbitrary, the sequence $\{u_n\}_{n\in\mathbb{N}}$ converges to 0 in the topology of $L^{p-}([0,1])$. $\square$

(v) $\mathrm{sim}\big(\mathcal{B}([0,1])\big)$ is dense in $L^{p-}([0,1])$.

Proof:

This follows immediately from (iii) and (iv) in combination with Lemma 2.3.2. $\square$

Further properties of the space $L^{p-}([0,1])$ may be found in [2]. ◀

Example 2.3.2
The Fréchet function space $L^p_{\mathrm{loc}}(\mathbb{R})$

Let $\Omega := \mathbb{R}$ and let λ be Lebesgue measure. Note that in this case $\big(\mathbb{R}, \mathcal{B}(\mathbb{R}), \lambda\big)$ is a σ-finite measure space. By $L^0(\mathbb{R})$ we denote the Lebesgue measurable functions $f : \mathbb{R} \to \mathbb{C}$. Fix $p \in [1, \infty)$. Define, for each $k \in \mathbb{N}$, a mapping $q_k : L^0(\mathbb{R}) \to [0, \infty]$ by

$$q_k(f) := \left(\int_{-k}^k |f|^p \, d\lambda \right)^{1/p}.$$

Then each q_k is a function semi-norm as for $f, g \in L^0(\mathbb{R})$ satisfying $|f| \leqslant |g|$ we have $|f|^p \leqslant |g|^p$ and, hence,

$$q_k(f) = \left(\int_{-k}^k |f|^p \, d\lambda \right)^{1/p} \leqslant \left(\int_{-k}^k |g|^p \, d\lambda \right)^{1/p} = q_k(g),$$

for $k \in \mathbb{N}$. As $[-k, k] \subseteq [-(k+1), k+1]$, for all $k \in \mathbb{N}$, it is clear that the sequence of function semi-norms $\{q_k\}_{k\in\mathbb{N}}$ is increasing, i.e., $q_k(f) \leqslant q_{k+1}(f)$, for all $f \in L^0(\mathbb{R})$. Again, the sequence is fundamental since for each $0 \neq f \in L^0(\mathbb{R})$ there exists at least one $m \in \mathbb{N}$ such that $q_m(f) \in (0, \infty]$. Thus,

$$L^p_{\mathrm{loc}}(\mathbb{R}) := \bigcap_{k\in\mathbb{N}} \{f \in L^0(\mathbb{R}) : q_k(f) < \infty\}$$

is a metrizable function space whose topology is generated by $\{q_k\}_{k\in\mathbb{N}}$. Since $L^p_{\mathrm{loc}}(\mathbb{R})$ is complete, [23, p. 40], it is a Fréchet function space. Some important properties of the space $L^p_{\mathrm{loc}}(\mathbb{R})$ are the following ones.

(i) $\left(L_{\mathrm{loc}}^p(\mathbb{R})\right)^* = \left\{ f \in L_{\mathrm{loc}}^q(\mathbb{R}) : f \text{ is compactly supported} \right\}$

Proof:

Let $\varphi : L_{\mathrm{loc}}^p(\mathbb{R}) \to \mathbb{C}$ be a continuous linear functional. Then there exist $k \in \mathbb{N}$ and $M > 0$ such that

$$|\langle f, \varphi \rangle| \leqslant M \, q_k(f) = M \left(\int_{-k}^k |f|^p \, d\lambda \right)^{1/p}, \qquad (2.25)$$

for all $f \in L_{\mathrm{loc}}^p(\mathbb{R})$. In particular,

$$|\langle f, \varphi \rangle| \leqslant M \left(\int_{-k}^k |f|^p \, d\lambda \right)^{1/p},$$

for all $f \in L^p([-k,k]) \subseteq L_{\mathrm{loc}}^p(\mathbb{R})|_{[-k,k]}$ where, for each $k \in \mathbb{N}$, the space $L^p([-k,k])$ is a Banach space. It is known that there then exists $g \in L^q([-k,k])$ (where q is the conjugate exponent of p) such that

$$\langle f, \varphi \rangle = \int_{-k}^k fg \, d\lambda, \quad \text{for all } f \in L^p([-k,k]). \qquad (2.26)$$

Define $\tilde{g} : \mathbb{R} \to \mathbb{C}$ by

$$\tilde{g}(w) := \begin{cases} g(w), & w \in [-k,k], \\ 0, & w \in \mathbb{R}\backslash[-k,k], \end{cases}$$

in which case $\tilde{g} \in L_{\mathrm{loc}}^q(\mathbb{R})$. Then it follows from (2.26) that, for each $f \in L_{\mathrm{loc}}^p(\mathbb{R})$,

$$\begin{aligned} \langle f, \varphi \rangle &= \left\langle f\chi_{[-k,k]} + f\chi_{\mathbb{R}\backslash[-k,k]}, \varphi \right\rangle \\ &= \left\langle f\chi_{[-k,k]}, \varphi \right\rangle + \left\langle f\chi_{\mathbb{R}\backslash[-k,k]}, \varphi \right\rangle \\ &= \int_{-k}^k fg \, d\lambda + \left\langle f\chi_{\mathbb{R}\backslash[-k,k]}, \varphi \right\rangle \end{aligned}$$

because $f\chi_{[-k,k]} \in L^p([-k,k])$. But, by (2.25) we have

$$\left| \left\langle f\chi_{\mathbb{R}\backslash[-k,k]}, \varphi \right\rangle \right| \leqslant M \left(\int_{-k}^k \left| f\chi_{\mathbb{R}\backslash[-k,k]} \right|^p d\lambda \right)^{1/p} = 0.$$

Accordingly,

$$\langle f, \varphi \rangle = \int_{\mathbb{R}} f\tilde{g} \, d\lambda, \quad \text{for all } f \in L_{\mathrm{loc}}^p(\mathbb{R}),$$

where $\tilde{g} \in L_{\mathrm{loc}}^q(\mathbb{R})$ is compactly supported.

Conversely, if $h \in L_{\text{loc}}^q(\mathbb{R})$ satisfies $h(w) = 0$ for all $w \notin K$ with $K \subseteq \mathbb{R}$ being compact, then for every $f \in L_{\text{loc}}^p(\mathbb{R})$ we have (via Hölder's inequality)

$$\int_{\mathbb{R}} fh \, d\lambda = \int_K fh \, d\lambda \overset{(2.9)}{\leqslant} \|f\|_{L^p(K)} \|h\|_{L^q(K)} \leqslant \|h\|_{L^q(K)} \, q_k(f),$$

where $k \in \mathbb{N}$ is chosen to satisfy $K \subseteq [-k, k]$. This shows that the linear functional

$$f \mapsto \int_{\mathbb{R}} fh \, d\lambda, \quad \text{for } f \in L_{\text{loc}}^p(\mathbb{R}),$$

is continuous on $L_{\text{loc}}^p(\mathbb{R})$. $\quad\square$

(ii) $L_{\text{loc}}^p(\mathbb{R})$ is reflexive for $1 < p < \infty$.

Proof:

For $1 < p < \infty$, each local Banach space $L^p([-k, k]) = L_{\text{loc}}^p(\mathbb{R})/q_k^{-1}(\{0\})$ is reflexive. Hence, by Proposition 2.1.2, $L_{\text{loc}}^p(\mathbb{R})$ is reflexive as well. $\quad\square$

(iii) $L^\infty(\mathbb{R}) \subseteq L_{\text{loc}}^p(\mathbb{R})$. In particular, $\text{sim}(\mathcal{B}(\mathbb{R})) \subseteq L_{\text{loc}}^p(\mathbb{R})$.

Proof:

Let $f \in L^\infty(\mathbb{R})$. Then $|f(w)| < M$, for λ-almost every $w \in \mathbb{R}$, for some real number $M > 0$. Thereby, for each $k \in \mathbb{N}$, we obtain

$$q_k(f) = \left(\int_{-k}^k |f(w)|^p \, d\lambda(w) \right)^{1/p} \leqslant \left(\int_{-k}^k M^p \, d\lambda(w) \right)^{1/p}$$

$$= M \, \lambda([-k, k])^{1/p} < \infty.$$

Hence, $L^\infty(\mathbb{R}) \subseteq L_{\text{loc}}^p(\mathbb{R})$. Since each $\mathcal{B}(\mathbb{R})$-simple function is bounded on $\mathbb{R}$, it follows that $\text{sim}(\mathcal{B}(\mathbb{R})) \subseteq L^\infty(\mathbb{R})$ and consequently, also $\text{sim}(\mathcal{B}(\mathbb{R})) \subseteq L_{\text{loc}}^p(\mathbb{R})$. $\square$

(iv) $L_{\text{loc}}^p(\mathbb{R})$ has a σ-Lebesgue topology.

Proof:

Let $\{u_n\}_{n \in \mathbb{N}} \subseteq L_{\text{loc}}^p(\mathbb{R})^+$ be any sequence of functions satisfying $u_n \downarrow_n 0$ pointwise on $\mathbb{R}$. Then $0 \leqslant u_n \leqslant u_1$, for all $n \in \mathbb{N}$. Fix $k \in \mathbb{N}$. Then also $0 \leqslant u_n^p \leqslant u_1^p$ with $u_n^p \downarrow_n 0$ pointwise. Since $u_1 \in L^p([-k, k])$ we have $u_1^p \in L^1([-k, k])$. It follows from Lebesgue's Dominated Convergence Theorem 2.2.2 that

$$\lim_{n \to \infty} q_k^p(u_n) = \lim_{n \to \infty} \int_{-k}^k |u_n|^p \, d\lambda = \int_{-k}^k \left(\lim_{n \to \infty} |u_n|^p \right) d\lambda = 0,$$

and hence, also $\lim_{n \to \infty} q_k(u_n) = 0$. Since $k \in \mathbb{N}$ is arbitrary, the sequence $\{u_n\}_{n \in \mathbb{N}}$ converges to 0 in the topology of $L^p_{\mathrm{loc}}(\mathbb{R})$. $\quad\square$

(v) $\mathrm{sim}\big(\mathcal{B}(\mathbb{R})\big)$ is dense in $L^p_{\mathrm{loc}}(\mathbb{R})$.

Proof:

This follows from (iii), (iv) and Lemma 2.3.2. $\quad\square\quad\blacktriangleleft$

2.4   Vector measures

Let (Ω, Σ) be a measurable space and let $(X, \{p_k\}_{k \in \mathbb{N}})$ be a Fréchet space with dual space X^*. A mapping $m : \Sigma \to X$ is called a *vector measure* if it is σ-additive, i.e., if

$$m \left(\bigcup_{j=1}^{\infty} A_j \right) = \sum_{j=1}^{\infty} m(A_j),$$

for any sequence $\{A_j\}_{j \in \mathbb{N}} \subseteq \Sigma$ of disjoint sets. Equivalently, if m is finitely additive, then m is σ-additive if and only if the sequence $\{m(A_j)\}_{j \in \mathbb{N}} \subseteq X$ converges to 0 in the topology of X whenever $\{A_j\}_{j \in \mathbb{N}} \subseteq \Sigma$ satisfies $A_j \downarrow_j \varnothing$ pointwise on Ω.

For each $x^* \in X^*$, define a $\mathbb{C}$-valued measure $\langle m, x^* \rangle : \Sigma \to \mathbb{C}$ by

$$\langle m, x^* \rangle(A) := \langle m(A), x^* \rangle, \quad \text{for } A \in \Sigma.$$

It was already noted that the variation measure $|\langle m, x^* \rangle|$ of each complex measure $\langle m, x^* \rangle$, for $x^* \in X^*$, is finite, [31, p. 144].

Remark 2.4.1

A finitely additive map $m : \Sigma \to X$ is σ-additive if and only if the $\mathbb{C}$-valued map $\langle m, x^* \rangle : A \mapsto \langle m(A), x^* \rangle$, for $A \in \Sigma$, is σ-additive for every $x^* \in X^*$.

Proof of Remark 2.4.1:

Suppose that $m : \Sigma \to X$ is σ-additive. Let $\{A_j\}_{j \in \mathbb{N}} \subseteq \Sigma$ be any sequence of pairwise disjoint sets. Since each $x^* \in X^*$ is continuous it follows that

$$
\begin{aligned}
\langle m, x^* \rangle \left(\bigcup_{j=1}^{\infty} A_j \right) &= \left\langle m \left(\bigcup_{j=1}^{\infty} A_j \right), x^* \right\rangle \\
&= \left\langle \sum_{j=1}^{\infty} m(A_j), x^* \right\rangle \\
&= \sum_{j=1}^{\infty} \langle m(A_j), x^* \rangle
\end{aligned}
$$

38

$$= \sum_{j=1}^{\infty} \langle m, x^* \rangle(A_j),$$

for all $x^* \in X^*$. Hence, $\langle m, x^* \rangle$ is σ-additive.

Conversely, suppose now that $\langle m, x^* \rangle$ is σ-additive, for each $x^* \in X^*$. Again, let $\{A_j\}_{j \in \mathbb{N}} \subseteq \Sigma$ be any sequence of pairwise disjoint sets. Consider any increasing sequence of natural numbers $\{j_k\}_{k \in \mathbb{N}} \subseteq \mathbb{N}$. Fix $x^* \in X^*$. Due to the σ-additivity of $\langle m, x^* \rangle$ we have

$$\langle m, x^* \rangle \left(\bigcup_{j=1}^{\infty} A_j \right) = \sum_{j=1}^{\infty} \langle m, x^* \rangle(A_j)$$

and therefore obtain that

$$
\begin{aligned}
\left\langle m \left(\bigcup_{k=1}^{\infty} A_{j_k} \right), x^* \right\rangle &= \langle m, x^* \rangle \left(\bigcup_{k=1}^{\infty} A_{j_k} \right) \\
&= \sum_{k=1}^{\infty} \langle m, x^* \rangle(A_{j_k}) \\
&= \sum_{k=1}^{\infty} \langle m(A_{j_k}), x^* \rangle \\
&= \left\langle \sum_{k=1}^{\infty} m(A_{j_k}), x^* \right\rangle.
\end{aligned}
$$

Since $x^* \in X^*$ is arbitrary, this means that the subseries $\sum_{k=1}^{\infty} m(A_{j_k})$ of $\sum_{j=1}^{\infty} m(A_j)$ is weakly convergent to $m(\bigcup_{k=1}^{\infty} A_{j_k})$. The Orlicz-Pettis Theorem 2.1.3 implies then that the series $\sum_{j=1}^{\infty} m(A_j)$ converges unconditionally to $m(\bigcup_{j=1}^{\infty} A_j)$ in the topology of X. Thus, m is σ-additive in X. $\square$

Let $m : \Sigma \to X$ be a vector measure. A measurable function $f : \Omega \to \mathbb{C}$ is called *scalarly m-integrable* if it is integrable with respect to each scalar measure $\langle m, x^* \rangle$, for $x^* \in X^*$, that is,

$$\int_{\Omega} |f| \, d|\langle m, x^* \rangle| < \infty.$$

A function $f : \Omega \to \mathbb{C}$ is said to be *m-integrable* if it is scalarly m-integrable and if, for each $A \in \Sigma$, there exists an element $\int_A f \, dm \in X$ such that

$$\left\langle \int_A f \, dm, x^* \right\rangle = \int_A f \, d\langle m, x^* \rangle, \quad \text{for all } x^* \in X^*.$$

The set function $m_f : \Sigma \to X$ defined by

$$m_f(A) := \int_A f \, dm, \quad \text{for } A \in \Sigma,$$

is called the *indefinite integral* of f with respect to m. By the Orlicz-Pettis Theorem 2.1.3 it is also a vector measure, [19, p. 160]. Note that each Σ-simple function $s : \Omega \to \mathbb{C}$ of the form (2.4) is m-integrable; its m-integral is defined by

$$\int_A s \, dm := \sum_{j=1}^{l} \alpha_j \, m(A \cap A_j). \tag{2.27}$$

The following alternative description of m-integrability of a function $f : \Omega \to \mathbb{C}$ is given in [19, pp. 161–162].

Proposition 2.4.1

Let X be a Fréchet space, $m : \Sigma \to X$ be a vector measure and $f : \Omega \to \mathbb{C}$ be a function. Then the following assertions are equivalent:

(i) f is m-integrable.

(ii) There exists a sequence $\{s_n\}_{n \in \mathbb{N}} \subseteq \mathrm{sim}(\Sigma)$ of Σ-simple functions which converges pointwise to f on Ω and such that, for each $A \in \Sigma$, the sequence $\{\int_A s_n \, dm\}_{n \in \mathbb{N}} \subseteq X$ converges in the topology of X.

In this case, $\int_A f \, dm = \lim_{n \to \infty} \int_A s_n \, dm$, for each $A \in \Sigma$. $\quad\square$

Let, for each $k \in \mathbb{N}$, $X/p_k^{-1}(\{0\})$ be the quotient space determined by the semi-norm p_k and denote by X_k its local Banach space with $\|\cdot\|_k$ being the norm in X_k. Denote by Π_k the canonical quotient map

$$\Pi_k : X \to X/p_k^{-1}(\{0\});$$

the same notation is used when Π_k is interpreted as being X_k-valued. It is clear that Π_k is continuous. Define, for each $k \in \mathbb{N}$, the set function

$$m_k := \Pi_k \circ m : \Sigma \to X/p_k^{-1}(\{0\}). \tag{2.28}$$

The continuity of Π_k ensures that m_k is a vector measure on Σ again, with values in $X/p_k^{-1}(\{0\}) \hookrightarrow X_k$. The *variation measure* $|m_k| : \Sigma \to [0, \infty]$ of the Banach space-valued vector measure m_k is then defined in the usual way:

$$|m_k|(A) := \sup_\pi \sum_{j=1}^l \|m_k(A_j)\|_k, \quad \text{for } A \in \Sigma,$$

where the supremum is taken over all finite partitions $\pi = \{A_j\}_{j=1}^l$ of A. The variation measure $|m_k|$, which is always σ-additive with values in $[0, \infty]$, is called *finite* if $|m_k|(\Omega) < \infty$. Moreover, $m : \Sigma \to X$ is said to have *finite variation* if $|m_k|$ is finite, for each $k \in \mathbb{N}$.

The space of all m-integrable functions is denoted by $\mathcal{L}^1(m)$. Note that, whenever $(X, \{p_k\}_{k \in \mathbb{N}})$ is a Fréchet space generated by a fundamental sequence of increasing semi-norms $\{p_k\}_{k \in \mathbb{N}}$, the sets

$$B_k := \big\{x \in X : p_k(x) \leqslant 1\big\}, \quad \text{for } k \in \mathbb{N},$$

form a fundamental sequence of zero neighbourhoods for X and their *polars*

$$B_k^\circ := \big\{x^* \in X^* : |\langle x, x^* \rangle| \leqslant 1, \text{ for all } x \in B_k\big\}, \quad \text{for } k \in \mathbb{N},$$

are absolutely convex, [22, p. 245]. Moreover, $\{B_k^\circ\}_{k \in \mathbb{N}}$ is a fundamental sequence of bounded sets in the strong dual X_β^*, i.e., each bounded set in X_β^* is contained in a multiple of B_k° for some $k \in \mathbb{N}$. In addition, each set B_k°, for $k \in \mathbb{N}$, is a Banach disc, that is, the linear hull

$$X_{B_k^\circ} := \bigcup_{\lambda > 0} \lambda B_k^\circ$$

generated by B_k° in X^* is a Banach space when equipped with its Minkowski functional

$$\phi_{B_k^\circ}(x^*) := \inf\big\{\lambda > 0 : x^* \in \lambda B_k^\circ\big\}, \quad \text{for } x^* \in X_{B_k^\circ},$$

[22, p. 278]. For $k \in \mathbb{N}$ fixed, the p_k-*semi-variation* of m is the set function $\tilde{p}_k(m) : \Sigma \to [0, \infty)$ given by

$$\tilde{p}_k(m)(A) := \sup\big\{|\langle m, x^* \rangle|(A) : x^* \in B_k^\circ\big\}, \quad \text{for } A \in \Sigma.$$

The following inequalities concerning the p_k-semi-variation of m are fundamental in the theory of vector measures. Respective results are found in [17, Lemma II.1.2], where there is 2 in place of 4 because X is considered over $\mathbb{R}$ rather than $\mathbb{C}$ and in [29, Proposition I.2] for the case that X is a Banach space.

41

Proposition 2.4.2

Let X be a Fréchet space and $m : \Sigma \to X$ be a vector measure. Then, for each $A \in \Sigma$, the inequalities

$$\sup\{p_k(m(B)) : B \in \Sigma, B \subseteq A\} \leqslant \tilde{p}_k(m)(A) \leqslant 4\sup\{p_k(m(B)) : B \in \Sigma, B \subseteq A\} \tag{2.29}$$

hold, for all $k \in \mathbb{N}$.

Proof:

Note that X^* determines the topology of X in that

$$p_k(x) = \sup\{|\langle x, x^* \rangle| : x^* \in B_k^\circ\}, \quad \text{for } x \in X, \tag{2.30}$$

for $k \in \mathbb{N}$, [22, §22]. Fix $k \in \mathbb{N}$ and choose an arbitrary $A \in \Sigma$. Let $B \in \Sigma$ such that $B \subseteq A$. Then (2.30) and Lemma 2.2.1 (i) and (ii) imply that

$$
\begin{aligned}
p_k(m(B)) &\overset{(2.30)}{=} \sup\{|\langle m(B), x^* \rangle| : x^* \in B_k^\circ\} \\
&\leqslant \sup\{|\langle m, x^* \rangle|(B) : x^* \in B_k^\circ\} \\
&\leqslant \sup\{|\langle m, x^* \rangle|(A) : x^* \in B_k^\circ\} \\
&= \tilde{p}_k(m)(A).
\end{aligned}
$$

Thus, also $\sup\{p_k(m(B)) : B \in \Sigma, B \subseteq A\} \leqslant \tilde{p}_k(m)(A)$ holds.

On the other hand, for any $x^* \in B_k^\circ$, we have

$$|\langle m(B), x^* \rangle| \leqslant \sup\{|\langle m(B), z^* \rangle| : z^* \in B_k^\circ\} \overset{(2.30)}{=} p_k(m(B)). \tag{2.31}$$

Hence, by Lemma 2.2.1 (iii), we obtain that

$$
\begin{aligned}
\tilde{p}_k(m)(A) &= \sup\{|\langle m, x^* \rangle|(A) : x^* \in B_k^\circ\} \\
&\leqslant 4\sup\{|\langle m(B), x^* \rangle| : B \in \Sigma, B \subseteq A\} \\
&\overset{(2.31)}{\leqslant} 4\sup\{p_k(m(B)) : B \in \Sigma, B \subseteq A\}.
\end{aligned}
$$

$A \in \Sigma$ and $k \in \mathbb{N}$ were arbitrary. Thus, both inequalities as stated in the assertion of Proposition 2.4.2 hold, for all $A \in \Sigma$, for all $k \in \mathbb{N}$. $\square$

Define then, for $f \in \mathcal{L}^1(m)$, its p_k-**upper integral** by

$$\tilde{p}_k(m)(f) := \tilde{p}_k(m_f)(\Omega) = \sup\{|\langle m_f, x^* \rangle|(\Omega) : x^* \in B_k^\circ\}.$$

The sequence of semi-norms $\{\tilde{p}_k(m)\}_{k \in \mathbb{N}}$ defines a topology on $\mathcal{L}^1(m)$ (see also p.

45). The inequality (2.29) implies that an equivalent topology on $\mathcal{L}^1(m)$ is generated by the semi-norms $p_k(m) : \mathcal{L}^1(m) \to [0, \infty)$ defined by

$$p_k(m)(f) := \sup\{p_k(m_f(A)) : A \in \Sigma\}, \quad \text{for } f \in \mathcal{L}^1(m),$$

for $k \in \mathbb{N}$.

An m-integrable function $f \in \mathcal{L}^1(m)$ is called m-*null* if its indefinite integral m_f is the zero vector measure, i.e., if $m_f(A) = 0$, for all $A \in \Sigma$. By definition of the semi-norms $p_k(m)$ this is equivalent to $p_k(m)(f) = 0$, for all $k \in \mathbb{N}$. Equivalently, a function $f \in \mathcal{L}^1(m)$ is m-null if and only if f is $|m_k|$-null, for all $k \in \mathbb{N}$, [24, pp. 212–214]. Two m-integrable functions $f, g \in \mathcal{L}^1(m)$ are equal m-*almost everywhere* (briefly: m-a.e.) if $|f - g|$ is m-null. Denote by $\mathcal{N}(m)$ the subspace of all m-null functions and by $L^1(m)$ the quotient space $\mathcal{L}^1(m)/\mathcal{N}(m)$. Finally, a set $A \in \Sigma$ is said to be m-*null* if its characteristic function χ_A is m-null. Equivalently, $A \in \Sigma$ is m-null if $\tilde{p}_k(m)(A) = 0$, for all $k \in \mathbb{N}$. In view of (2.29) this is equivalent to $m(B \cap A) = 0$, for every $B \in \Sigma$. The family of all m-null sets is denoted by $\mathcal{N}_0(m)$. In the following remark, members of Σ are freely identified with their characteristic functions.

Remark 2.4.2
A function $f \in \mathcal{L}^1(m)$ is m-null if and only if $f^{-1}(\mathbb{C}\backslash\{0\})$ is an m-null set.

Proof of Remark 2.4.2:

Let $f \in \mathcal{L}^1(m)$ and define $B := f^{-1}(\mathbb{C}\backslash\{0\}) = \{w \in \mathbb{C} : f(w) \neq 0\}$. Then $f = 0$ on B^c and consequently

$$\int_{A \cap B^c} f \, dm = \int_{A \cap B^c} 0 \, dm = 0, \tag{2.32}$$

for all $A \in \Sigma$. By making use of the linearity of the integral, [19, p. 160], we obtain, for all $A \in \Sigma$, that

$$\int_A f \, dm = \int_A f\chi_\Omega \, dm = \int_A f\chi_{B \cup B^c} \, dm = \int_A f\chi_B \, dm + \int_A f\chi_{B^c} \, dm \overset{(2.32)}{=} \int_{A \cap B} f \, dm. \tag{2.33}$$

Let now f be an m-null function. Then f is also an $|m_k|$-null function, for all $k \in \mathbb{N}$. For $k \in \mathbb{N}$ fixed, this means that

$$\int_\Omega |f| \, d|m_k| = 0,$$

[24, p. 213]. Accordingly, $B := \{w \in \Omega : |f(w)| > 0\} = \{w \in \Omega : |f(w)| \neq 0\} =$

$f^{-1}(\mathbb{C}\backslash\{0\})$ is an $|m_k|$-null set. Due to the obvious inequality

$$0 \leqslant \|m_k(A)\|_k \leqslant |m_k|(A), \quad \text{for all } A \in \Sigma,$$

(where $\|\cdot\|_k$ is the norm in X_k) the set B is also an m_k-null set. But, $m_k = \Pi_k \circ m$, and it follows, for given $A \in \Sigma$, that we obtain the equality

$$m_k(A \cap B) = \Pi_k\big(m(A \cap B)\big) = 0.$$

Consequently, $m(A \cap B) \in p_k^{-1}(\{0\})$, meaning that

$$\|m(A \cap B)\|_k = p_k\big(m(A \cap B)\big) = 0.$$

As $A \in \Sigma$ and $k \in \mathbb{N}$ were arbitrary, we obtain that $m(A \cap B) = 0 \in X$, for all $A \in \Sigma$. Thus, $B = f^{-1}(\mathbb{C}\backslash\{0\})$ is indeed an m-null set.

Conversely, let B be an m-null set. Then $m(A \cap B) = 0$, for all $A \in \Sigma$. Thus, we have

$$\int_A f \, dm \overset{(2.33)}{=} \int_{A \cap B} f \, dm = 0,$$

for all $A \in \Sigma$, implying that $m_f(A) = 0$, for all $A \in \Sigma$, and consequently that f is an m-null function. $\quad\square$

Let X be a Fréchet space and $m : \Sigma \to X$ be a vector measure. A finite, positive measure $\nu : \Sigma \to [0, \infty)$ is called a *control measure* for m if the ν-null sets and the m-null sets coincide, i.e., if $\mathcal{N}_0(\nu) = \mathcal{N}_0(m)$.

Remark 2.4.3

Note that also a σ-finite measure $\nu : \Sigma \to [0, \infty]$ satisfying $\mathcal{N}_0(\nu) = \mathcal{N}_0(m)$ may be considered as a control measure for m since it is always possible to construct a finite measure $\tilde{\nu}$ out of ν which has the same null sets.

Proof of Remark 2.4.3:

Let $m : \Sigma \to X$ be a Fréchet-space-valued vector measure und let $\nu : \Sigma \to [0, \infty]$ be a σ-finite measure satisfying $\mathcal{N}_0(\nu) = \mathcal{N}_0(m)$. Choose any disjoint sequence $\{A_j\}_{j \in \mathbb{N}} \subseteq \Sigma$ satisfying $\bigcup_{j \in \mathbb{N}} A_j = \Omega$ and $\nu(A_j) < \infty$, for all $j \in \mathbb{N}$, and define $\tilde{\nu} : \Sigma \to [0, \infty)$ by

$$\tilde{\nu}(A) = \sum_{j=1}^{\infty} \frac{\nu(A \cap A_j)}{2^j (1 + \nu(A \cap A_j))}, \quad \text{for } A \in \Sigma.$$

44

Since for all $j \in \mathbb{N}$ we have $\nu(\Omega \cap A_j) < \infty$ and

$$\frac{\nu(\Omega \cap A_j)}{2^j(1 + \nu(\Omega \cap A_j))} < \frac{1}{2^j},$$

it is clear that $\tilde{\nu}(\Omega) < \infty$.

Furthermore, by definition of $\tilde{\nu}$ it is obvious that any ν-null set $A \in \Sigma$ becomes a $\tilde{\nu}$-null set, i.e., $\mathcal{N}_0(\nu) \subseteq \mathcal{N}_0(\tilde{\nu})$.

On the other hand, let $A \in \Sigma$ be a $\tilde{\nu}$-null set meaning that $\tilde{\nu}(A) = 0$. Then we have

$$\tilde{\nu}(A) = \sum_{j=1}^{\infty} \frac{\nu(A \cap A_j)}{2^j(1 + \nu(A \cap A_j))} = 0.$$

But this is equivalent to $\nu(A \cap A_j) = 0$, for all $j \in \mathbb{N}$. The σ-additivity of ν yields then that

$$\nu(A) = \nu(A \cap \Omega) = \nu\left(A \cap \bigcup_{j=1}^{\infty} A_j\right) = \sum_{j=1}^{\infty} \nu(A \cap A_j) = \sum_{j=1}^{\infty} 0 = 0.$$

So, A is also a ν-null set. It follows that $\mathcal{N}_0(\tilde{\nu}) \subseteq \mathcal{N}_0(\nu)$ and thus, the assertions of Remark 2.4.3 hold. $\square$

Let $(X, \{p_k\}_{k \in \mathbb{N}})$ be a Fréchet space and $m : \Sigma \to X$ be a vector measure. Then the space $L^1(m)$ becomes a *Fréchet function space* when equipped with the semi-norms $\{\tilde{p}_k(m)\}_{k \in \mathbb{N}}$. To see this, fix $k \in \mathbb{N}$ and consider the Banach-space-valued vector measure $m_k : \Sigma \to X_k$ given by (2.28). Denote by $\Pi_k^* : X_k^* \to \mathrm{Lin}(B_k^\circ) := \bigcup_{\lambda > 0} \lambda B_k^\circ$ the dual map of $\Pi_k : X \to X_k$. Here, X_k^* denotes the dual space of X_k. Note that Π_k^* is an isometric bijection, [23, Remark 24.5(b)]. Rybakov's Theorem, [8, p. 268], ensures that there exists an element $\zeta_k^* \in X_k^*$ such that $|\langle m_k, \zeta_k^* \rangle|$ is a control measure for m_k. Let $x_k^* := \Pi_k^*(\zeta_k^*)$ and define a finite, positive measure $\nu : \Sigma \to [0, \infty)$ by

$$\nu(A) := \sum_{k=1}^{\infty} \frac{|\langle m, x_k^* \rangle|(A)}{2^k(1 + |\langle m, x_k^* \rangle|(\Omega))}, \quad \text{for } A \in \Sigma.$$

Then ν is a control measure for m and $M(\nu) = M(m)$, [5, Proof of Theorem 2.5]. Define, for each $k \in \mathbb{N}$, a mapping $\tilde{\rho}_k : M(\nu) \to [0, \infty]$ by

$$\tilde{\rho}_k(f) := \sup\left\{\int_{\Omega} |f| \, d|\langle m, x^* \rangle| : x^* \in B_k^\circ\right\}, \quad \text{for } f \in M(\nu).$$

Then $\{\tilde{\rho}_k\}_{k \in \mathbb{N}}$ is an increasing sequence of function semi-norms on $M(\nu)$ and $L_{\{\tilde{\rho}_k\}} = L^1(m)$, [5, pp. 643–644]. Furthermore, $\tilde{\rho}_k|_{L^1(m)} = \tilde{p}_k(m)$, for all $k \in \mathbb{N}$, and $L^1(m)$

is a complete metrizable topological vector space when equipped with the topology generated by $\{\tilde{p}_k(m)\}_{k\in\mathbb{N}}$, [5, Proof of Theorem 2.5], hence, a Fréchet function space. Moreover, $L^1(m)$ has a σ-Lebesgue topology, [5, p. 644].

Note that if, for $k \in \mathbb{N}$ fixed, $\mathcal{L}^1(m_k)$ denotes the space of all m_k-integrable functions and $\mathcal{L}^1(|m_k|)$ the space of all $|m_k|$-integrable functions, the inclusions

$$\left(\bigcap_{k\in\mathbb{N}} \mathcal{L}^1(|m_k|) \right) / \mathcal{N}(m) \subseteq L^1(m) \subseteq \left(\bigcap_{k\in\mathbb{N}} \mathcal{L}^1(m_k) \right) / \mathcal{N}(m) \qquad (2.34)$$

hold, [24, pp. 213–214], with both inclusions being continuous when $\bigcap_{k\in\mathbb{N}} \mathcal{L}^1(|m_k|)$ is equipped with the topology generated by the increasing sequence of semi-norms

$$\rho_k(f) := \int_\Omega |f|\, d|m_k|, \quad \text{for } f \in \bigcap_{k\in\mathbb{N}} \mathcal{L}^1(|m_k|),$$

and $\bigcap_{k\in\mathbb{N}} \mathcal{L}^1(m_k)$ is equipped with the topology generated by the increasing sequence of semi-norms

$$\| \cdot \|_k(m_k)(f) := \sup \left\{ \left\| \int_A f\, dm_k \right\|_k : A \in \Sigma \right\}, \quad \text{for } f \in \bigcap_{k\in\mathbb{N}} \mathcal{L}^1(m_k).$$

Denote by $L_w^1(m)$ the space of all scalarly m-integrable functions $f : \Omega \to \mathbb{C}$. A measurable function $f : \Omega \to \mathbb{C}$ is an element of $L_w^1(m)$ if and only if

$$\sup \left\{ \int_\Omega |f|\, d|\langle m, x^* \rangle| : x^* \in B_k^\circ \right\} < \infty, \quad \text{for all } k \in \mathbb{N}, \qquad (2.35)$$

[5, pp. 642–643]. By the definition of m-integrability it is clear that $L^1(m) \subseteq L_w^1(m)$. In some cases, however, these two spaces coincide. In [17] and [20], for instance, it is shown that in a weakly sequentially complete Fréchet space X scalar m-integrability implies m-integrability, meaning that in this case also $L_w^1(m) \subseteq L^1(m)$ is true. In particular, this is the case if X is reflexive, [5, p. 643]. Further properties of the space $L_w^1(m)$ have been investigated in [5], [7].

For the following definitions, let $\mu : \Sigma \to [0, \infty)$ be a finite measure. Note that the definition of a Σ-simple function S is still along the lines of (2.4) when S takes its values in a Fréchet space X instead of the complex numbers $\mathbb{C}$, i.e., S is of the form

$$S = \sum_{j=1}^{l} x_j \chi_{A_j}$$

where $x_1, \ldots, x_l \in X$ and $A_j := S^{-1}(\{x_j\})$, for each $j = 1, \ldots, l$, and $\bigcup_{j=1}^{l} A_j = \Omega$. The *Bochner μ-integral* of S is then defined in the obvious way by

$$(\mathrm{B}) - \int_{\Omega} S \, d\mu := \sum_{j=1}^{l} x_j \, \mu(A_j).$$

A function $H : \Omega \to X$ is called *strongly μ-measurable* if there exists a sequence of Σ-simple functions $H_n : \Omega \to X$, for $n \in \mathbb{N}$, such that for μ-almost every $w \in \Omega$

$$\lim_{n \to \infty} p_k \big(H_n(w) - H(w) \big) = 0, \quad \text{for all } k \in \mathbb{N}.$$

A strongly μ-measurable function $H : \Omega \to X$ is called *Bochner μ-integrable* if for each $w \in \Omega$,

$$\int_{\Omega} p_k \big(H(w) \big) \, d\mu < \infty, \quad \text{for all } k \in \mathbb{N}.$$

The following equivalent formulation, which combines both properties, is given in [24, pp. 214–215].

Lemma 2.4.1
A function $H : \Omega \to X$ is Bochner μ-integrable if and only if there exists a sequence of Σ-simple functions $H_n : \Omega \to X$, for $n \in \mathbb{N}$, such that

(i) $\lim_{n \to \infty} p_k \big(H_n(w) - H(w) \big) = 0$, *for all $k \in \mathbb{N}$, for μ-almost every $w \in \Omega$.*

(ii) $\lim_{n \to \infty} \int_{\Omega} p_k \big(H_n(w) - H(w) \big) \, d\mu = 0$, *for all $k \in \mathbb{N}$.* $\square$

Note, for a Bochner μ-integrable function H, that the Bochner μ-integral of H over A is then defined by

$$(\mathrm{B}) - \int_{A} H \, d\mu = \lim_{n \to \infty} (\mathrm{B}) - \int_{A} H_n \, d\mu.$$

This definition is independent of the choice of the sequence $\{H_n\}_{n \in \mathbb{N}}$. The indefinite Bochner μ-integral $\mu_H : \Sigma \to X$ of a Bochner μ-integrable function H is given by

$$\mu_H(A) := (\mathrm{B}) - \int_{A} H \, d\mu, \quad \text{for } A \in \Sigma, \tag{2.36}$$

and satisfies

$$\left\langle (\mathrm{B}) - \int_{A} H \, d\mu, x^* \right\rangle = \int_{A} \langle H(w), x^* \rangle \, d\mu(w), \quad \text{for } x^* \in X^*.$$

It is a vector measure of finite variation, [24, p. 216], where, for each $k \in \mathbb{N}$, the variation measure $|(\mu_H)_k| : \Sigma \to [0, \infty)$ is given by

$$|(\mu_H)_k|(A) = \int_A p_k(H(w)) \, d\mu, \quad \text{for } A \in \Sigma.$$

A third definition concerning the Bochner μ-integrability of a Fréchet-space-valued function $H : \Omega \to X$ is stated in [34, p. 75]. Namely, a function $H : \Omega \to X$ is said to be Bochner μ-integrable if there exists a Banach space $X_B \hookrightarrow X$ and a set $\Omega_0 \in \Sigma$ with $\mu(\Omega \backslash \Omega_0) = 0$ such that $H(w) \in X_B$, for $w \in \Omega_0$, and such that $H : \Omega_0 \to X_B$ is Bochner μ-integrable as a Banach-space-valued function.

A function $H : \Omega \to X$ is called *weakly μ-measurable* if the $\mathbb{C}$-valued function

$$w \mapsto \langle H(w), x^* \rangle, \quad \text{for } w \in \Omega,$$

is Σ-measurable, for each $x^* \in X^*$. A weakly μ-measurable function H is said to be *Pettis μ-integrable* if

$$\int_\Omega |\langle H, x^* \rangle| \, d\mu < \infty, \tag{2.37}$$

for each $x^* \in X^*$, and if, for each $A \in \Sigma$, there exists an element $\int_A H \, d\mu \in X$ satisfying

$$\left\langle \int_A H \, d\mu, x^* \right\rangle = \int_A \langle H, x^* \rangle \, d\mu.$$

If X is separable and reflexive, then a weakly μ-measurable function $H : \Omega \to X$ is Pettis μ-integrable if and only if it satisfies (2.37) for every $x^* \in X^*$, [34, Corollary 4.1].

2.5   Integration on topological groups

A non-empty set G endowed with a function (also called "operation")

$$\star : G \times G \to G, \quad (x, y) \mapsto x \star y$$

is called a *group* and denoted by $(G, \star)$ if it satisfies the following conditions:

   (i) $(x \star y) \star z = x \star (y \star z)$, for all $x, y, z \in G$.
   (ii)   There is a unique element $e \in G$ such that $x \star e = x = e \star x$, for all $x \in G$.
   (iii) For each $x \in G$, there is a unique $x^{-1} \in G$ such that $x \star x^{-1} = e = x^{-1} \star x$.

A group is called *Abelian* or *commutative* if it additionally satisfies

   (iv) $x \star y = y \star x$, for all $x, y \in G$.

Let τ be a topology on G. The triple $(G, \star, \tau)$ is called a *topological group* if $(G, \star)$

is a group and (G, τ) is a topological space such that both the group operation $\star$ and the inverse function

$$\cdot^{-1} : G \to G, \quad x \mapsto x^{-1}$$

are continuous with respect to the product topology on $G \times G$ resp. the given topology τ on G. A topological group is called *locally compact* if (G, τ) is a locally compact topological space.

In the sequel, let G be a compact Abelian group and denote by $C(G)$ the vector space of all continuous, $\mathbb{R}$-valued functions on G. As usual, $C(G)^+$ denotes the non-negative functions in $C(G)$. The group operation $\star$ is written as $+$ in this case, i.e., $x + y$ in place of $x \star y$.

On G there exists an invariant integral, meaning that there exist a translation invariant, positive linear functional I defined on $C(G)$ and, associated with I, a translation invariant, finite, positive measure μ. Writing

$$I(f) = \int_G f \, d\mu = \int_G f(x) \, d\mu(x)$$

this means, for all $f, g \in C(G)$ and $\lambda \in \mathbb{R}$, that the following conditions hold:

(i) $\int_G (f + g) \, d\mu = \int_G f \, d\mu + \int_G g \, d\mu$.

(ii) $\int_G (\lambda f) \, d\mu = \lambda \int_G f \, d\mu$.

(iii) $\int_G f \, d\mu \geqslant 0$, if $f \in C(G)^+$.

(iv) $\int_G f \, d\mu > 0$, if $f \in C(G)^+$ with $f \neq 0$.

(v) $\int_G f(x + y) \, d\mu(x) = \int_G f(x) \, d\mu(x)$, for all $y \in G$.   (translation invariance)

(vi) If $f, g \in C(G)$ satisfy $f \leqslant g$, then $\int_G f \, d\mu \leqslant \int_G g \, d\mu$.

(vii) $\left| \int_G f \, d\mu \right| \leqslant \int_G |f| \, d\mu$.

The integral is then gradually extended to all those complex-valued functions $f : G \to \mathbb{C}$ that are integrable with respect to the measure μ, [27, p. 234 & p. 282]. The integral, which is unique up to a multiplicative positive constant, is called the *Haar integral* and the associated measure is called *Haar measure*. It is possible to choose μ (which we do) such that $\mu(G) = 1$.

In accordance to the terminology used in classical measure theory, the space of all functions integrable with respect to the Haar measure μ will be denoted by $L^1(G)$ and the space of all functions p-integrable with respect to μ, $1 < p < \infty$, is denoted by $L^p(G)$ respectively. Note that each space $L^p(G)$, for $1 \leqslant p < \infty$, is a Banach space when equipped with the norm

$$\|f\|_p := \left(\int_G |f|^p \, d\mu \right)^{1/p}.$$

Define for functions $f, g \in L^1(G)$ the *convolution* $f * g : G \to \mathbb{C}$ by

$$(f * g)(x) := \int_G f(y) \, g(x - y) \, d\mu(x) = \int_G f(x + y) \, g(-y) \, d\mu(x).$$

Note that $L^1(G)$ endowed with the convolution as multiplication forms a Banach algebra, [27, pp. 288–289], i.e.,

$$\|f * g\|_1 \leqslant \|f\|_1 \|g\|_1, \quad \text{for all } f, g \in L^1(G).$$

More generally, for $f \in L^1(G)$ and $g \in L^p(G)$, $1 \leqslant p \leqslant \infty$, we have $f * g \in L^p(G)$ and

$$\|f * g\|_p \leqslant \|f\|_1 \|g\|_p, \tag{2.38}$$

[27, p. 288]. Furthermore, $L^1(G)$ is commutative, i.e.,

$$f * g = g * f, \quad \text{for all } f, g \in L^1(G),$$

as we are assuming that G is Abelian, [27, p. 289].

Let G be a compact Abelian group and consider the one-dimensional circle group

$$\mathbb{T} := S^1 := \{z \in \mathbb{C} : |z| = 1\},$$

where the group operation is multiplication in $\mathbb{C}$. A continuous homomorphism $\gamma : G \to \mathbb{T}$, i.e.,

$$\gamma(x + y) = \gamma(x) \, \gamma(y), \quad \text{for } x, y \in G,$$

is called a *character* of G. Endowed with the multiplication

$$(\gamma_1 \cdot \gamma_2)(x) = \gamma_1(x) \, \gamma_2(x), \quad \text{for } x \in G,$$

the set of all characters on G becomes an Abelian group and is denoted by $\hat{G}$, [27, p. 300]. It is called the *character group* of G. Note that the neutral element of $\hat{G}$ is the constant function 1, and for each $\gamma \in \hat{G}$ the inverse is the complex-conjugated function

$$\overline{\gamma} : x \mapsto \overline{\gamma(x)}, \quad \text{for } x \in G.$$

Equipped with the topology of compact convergence on G (that is, the topology of

50

uniform convergence on G) the group $(\hat{G}, \cdot)$ becomes an Abelian Hausdorff topological group, [27, p. 302], and is called the *dual group* of G. Since G is compact, $\hat{G}$ is discrete, [27, p. 303].

Finally, still with $(G, +)$ an (additive) compact Abelian group, let $L^1(G)$ be as defined before. Define, for $f \in L^1(G)$, the *Fourier transform* $\hat{f} : \hat{G} \to \mathbb{C}$ by

$$\hat{f}(\gamma) := \int_G f(x)\langle -x, \gamma \rangle \, d\mu(x), \quad \text{for } \gamma \in \hat{G}, \tag{2.39}$$

where $\langle \cdot, \cdot \rangle$ denotes the duality of the groups G and $\hat{G}$, i.e.,

$$\langle x, \gamma \rangle := \gamma(x), \quad \text{for } x \in G, \gamma \in \hat{G}.$$

Further definitions will be given in the Chapters 3 and 4 whenever there is a need for it.

Chapter 3

The optimal domain and integral extension of the operator T

The aim of Chapter 3 is to investigate the integration operator

$$I_{m_T} : L^1(m_T) \to X$$

associated with a Fréchet-space-valued vector measure $m_T : \Sigma \to X$ defined by

$$m_T(A) := T(\chi_A), \quad \text{for } A \in \Sigma,$$

where $T : X(\mu) \to X$ is a continuous linear operator defined on a Fréchet function space $X(\mu)$ over a σ-finite measure space (Ω, Σ, μ) and taking its values in the Fréchet space X. The main goal is to prove that $L^1(m_T)$ is the optimal domain of I_{m_T} (in a certain sense) when considered as continuous extension of the operator T to the "larger" domain $L^1(m_T)$. The respective investigations for $X(\mu)$ being a Banach function space and X being a Banach space have been exposed in [26]; see also the references there. So, the interesting part will be to see how the results differ when the problem is considered under these altered conditions. In Section 3.1 we prove the continuity of the inclusion maps

$$i : X(\mu) \to M(\mu) \quad \text{and} \quad j : X(\mu) \to Y(\mu)$$

where $X(\mu)$ and $Y(\mu)$ are two Fréchet function spaces over (Ω, Σ, μ) satisfying $X(\mu) \subseteq Y(\mu)$ as complex vector lattices. The continuity of the inclusion maps will be of importance when investigating the optimal domain of the operator T and its optimal extension. In Section 3.2 we turn our attention to the vector measure m_T as defined above. It turns out that the σ-Lebesgue topology of the Fréchet function space $X(\mu)$ and the μ-determinedness of the operator T play a crucial role for the theory. Section 3.3 presents the main result of this chapter. It states that $L^1(m_T)$

is the largest Fréchet function space having σ-Lebesgue topology into which $X(\mu)$ is continuously embedded and to which T admits an X-valued continuous linear extension. Furthermore, it is shown that such an extension is unique and is precisely the integration operator I_{m_T}.

3.1   The natural inclusion map $j : X(\mu) \to Y(\mu)$

In this section let (Ω, Σ, μ) be a σ-finite measure space. It will be the aim of this section to prove on the one hand the continuity of the inclusion map

$$i : X(\mu) \to M(\mu)$$

and to investigate on the other hand the inclusion map

$$j : X(\mu) \to Y(\mu)$$

whenever $X(\mu)$ and $Y(\mu)$ are two Fréchet function spaces satisfying $X(\mu) \subseteq Y(\mu)$ as complex vector sublattices.

We begin with two Lemmas which present a condition, that is necessary and sufficient for the completeness of the space $L_{\{q_k\}}$. The first Lemma is an unpublished result due to R. del Campo and W.J. Ricker, [6, Lemma 3.6].

Lemma 3.1.1

Let $\{q_k\}_{k \in \mathbb{N}}$ be an increasing fundamental sequence of function semi-norms. If the metrizable function space $L_{\{q_k\}}$ has the (JRF)-property, then

$$q_k \left(\sum_{n=1}^{\infty} |f_n| \right) \leqslant \sum_{n=1}^{\infty} q_k(f_n), \quad \text{for all } k \in \mathbb{N}, \tag{3.1}$$

for every sequence $\{f_n\}_{n \in \mathbb{N}} \subseteq L_{\{q_k\}}$ which is absolutely summable in $L_{\{q_k\}}$.

Proof:

Suppose that $L_{\{q_k\}}$ has the (JRF)-property. In order to establish (3.1), assume that there exists a sequence $\{f_n\}_{n \in \mathbb{N}} \subseteq L_{\{q_k\}}$ satisfying $\sum_{n=1}^{\infty} q_k(f_n) < \infty$, for all $k \in \mathbb{N}$, but, for some $m \in \mathbb{N}$, we have

$$q_m \left(\sum_{n=1}^{\infty} |f_n| \right) > \sum_{n=1}^{\infty} q_m(f_n).$$

54

Choose $\varepsilon > 0$ such that

$$q_m\left(\sum_{n=1}^{\infty}|f_n|\right) > \varepsilon + \sum_{n=1}^{\infty} q_m(f_n).$$

Observe, by the triangle inequality for q_m, that also

$$q_m\left(\sum_{n=j}^{\infty}|f_n|\right) > \varepsilon + \sum_{n=j}^{\infty} q_m(f_n), \quad \text{for all } j \in \mathbb{N}. \tag{3.2}$$

Since by assumption $\sum_{n=1}^{\infty} q_1(f_n) < \infty$, there exists $j_{1,1} \in \mathbb{N}$ with $\sum_{n \geqslant j_{1,1}} q_1(f_n) < 1^{-3}$. Because of $\sum_{n \geqslant j_{1,1}} q_1(f_n) < \infty$, we can find $j_{1,2} > j_{1,1}$ such that $\sum_{n \geqslant j_{1,2}} q_1(f_n) < 2^{-3}$. Proceed inductively to produce a strictly increasing sequence $\{j_{1,l}\}_{l \in \mathbb{N}} \subseteq \mathbb{N}$ satisfying

$$\sum_{n \geqslant j_{1,l}} q_1(f_n) < l^{-3}, \quad \text{for all } l \in \mathbb{N}.$$

Since $\sum_{n \geqslant j_{1,1}} q_2(f_n) < \infty$, there exists $j_{2,1} > j_{1,1}$ such that $\sum_{n \geqslant j_{2,1}} q_2(f_n) < 1^{-3}$. Because of $\sum_{n \geqslant \max\{j_{2,1},j_{1,2}\}} q_2(f_n) < \infty$, we can choose $j_{2,2} > \max\{j_{2,1}, j_{1,2}\}$ such that $\sum_{n \geqslant j_{2,2}} q_2(f_n) < 2^{-3}$. Assume that $j_{2,l-1}$ is already constructed for an arbitrary $l > 1$. Since $\sum_{n \geqslant \max\{j_{2,l-1},j_{1,l}\}} q_2(f_n) < \infty$, there exists $j_{2,l} \in \mathbb{N}$ satisfying $j_{2,l} > j_{2,l-1}$ and $j_{2,l} > j_{1,l}$ with $\sum_{n \geqslant j_{2,l}} q_2(f_n) < l^{-3}$. Thus, $\{j_{2,l}\}_{l \in \mathbb{N}}$ is a strictly increasing sequence satisfying $j_{2,l} > j_{1,l}$, for all $l \in \mathbb{N}$, and

$$\sum_{n \geqslant j_{2,l}} q_2(f_n) < l^{-3}, \quad \text{for all } l \in \mathbb{N}.$$

Continue inductively to produce for each $k \in \mathbb{N}$ a strictly increasing sequence $\{j_{k,l}\}_{l \in \mathbb{N}} \subseteq \mathbb{N}$ satisfying $j_{k+1,l} > j_{k,l}$ and

$$\sum_{n \geqslant j_{k,l}} q_k(f_n) < l^{-3}, \quad \text{for all } l \in \mathbb{N}.$$

Therefore, the diagonal sequence $\{j_l\}_{l \in \mathbb{N}}$ defined by $j_l := j_{l,l}$, for each $l \in \mathbb{N}$, is also strictly increasing. Moreover, for each $k \in \mathbb{N}$ we have $j_l = j_{l,l} \geqslant j_{k,l}$, for all $l \geqslant k$, and hence,

$$\sum_{n \geqslant j_l} q_k(f_n) \leqslant \sum_{n \geqslant j_{k,l}} q_k(f_n) < l^{-3}, \quad \text{for all } l \geqslant k. \tag{3.3}$$

Let $g_l := l \sum_{n=j_l}^{j_{l+1}} |f_n|$ and $g := \sum_{l=1}^{\infty} g_l$. On the one hand, for each $k \in \mathbb{N}$ we have

$$\sum_{l \geqslant k} q_k(g_l) \leqslant \sum_{l \geqslant k} l \sum_{n=j_l}^{j_{l+1}} q_k(f_n) \leqslant \sum_{l \geqslant k} l \sum_{n \geqslant j_l} q_k(f_n) \overset{(3.3)}{<} \sum_{l \geqslant k} l^{-2} < \infty.$$

Therefore, $\sum_{l=1}^{\infty} q_k(g_l) < \infty$, for all $k \in \mathbb{N}$, meaning that $\{g_l\}_{l \in \mathbb{N}}$ is an absolutely summable sequence in $L_{\{q_k\}}$. Since $L_{\{q_k\}}$ has the (JRF)-property, we can conclude that $q_k(g) < \infty$, for all $k \in \mathbb{N}$. On the other hand, for each $l \in \mathbb{N}$ we also have (pointwise on Ω) that

$$l \sum_{n \geqslant j_l} |f_n| \leqslant l \sum_{p \geqslant l} \sum_{n=j_p}^{j_{p+1}} |f_n| = \sum_{p \geqslant l} l \sum_{n=j_p}^{j_{p+1}} |f_n| \leqslant \sum_{p \geqslant l} p \sum_{n=j_p}^{j_{p+1}} |f_n| = \sum_{p \geqslant l} g_p \leqslant g. \qquad (3.4)$$

As q_k is a function semi-norm for each $k \in \mathbb{N}$ it follows that

$$q_m(g) \overset{(3.4)}{\geqslant} l\, q_m \left(\sum_{n \geqslant j_l} |f_n| \right) \overset{(3.2)}{>} l \left(\varepsilon + \sum_{n \geqslant j_l} q_m(f_n) \right) \geqslant l\varepsilon, \quad \text{for all } l \in \mathbb{N}.$$

Letting $l \to \infty$ we conclude that $q_m(g) = \infty$ which is a contradiction to our earlier conclusion that $q_k(g) < \infty$, for all $k \in \mathbb{N}$. $\qquad \square$

The next lemma is an extension of a result of Zaanen, who proved it for L_ρ being the space generated by a single function norm ρ, [36, p. 445]. For $L_{\{q_k\}}$ consisting of only $\mathbb{R}$-valued functions, see [6, Theorem 3.7]. We extend it to the setting of $\mathbb{C}$-valued functions.

Lemma 3.1.2

Let $L_{\{q_k\}}$ be the metrizable function space generated by an increasing fundamental sequence of function semi-norms $\{q_k\}_{k \in \mathbb{N}}$. Then the following assertions are equivalent:

(i) $L_{\{q_k\}}$ is complete.

(ii) $L_{\{q_k\}}$ has the (JRF)-property.

Proof:

(ii) $\Rightarrow$ (i) Assume that $L_{\{q_k\}}$ has the (JRF)-property. Let $\{f_n\}_{n \in \mathbb{N}} \subseteq L_{\{q_k\}}$ be a Cauchy sequence in $L_{\{q_k\}}$, meaning that for each $k \in \mathbb{N}$ and for each $\varepsilon > 0$ there exists an index $n_0(\varepsilon, k) \in \mathbb{N}$ such that $q_k(f_m - f_n) < \varepsilon$, for all $m, n \geqslant n_0(\varepsilon, k)$.

Hence, there exists an index $j_{1,1} \in \mathbb{N}$ such that $q_1(f_m - f_n) < 2^{-1}$, for all $m, n \geqslant j_{1,1}$. As $\{f_n\}_{n \geqslant j_{1,1}}$ is still Cauchy, we can find an index $j_{1,2} > j_{1,1}$ such that $q_1(f_m - f_n) < 2^{-2}$, for all $m, n \geqslant j_{1,2}$. Continuing this way we obtain a strictly

56

increasing sequence $\{j_{1,l}\}_{l\in\mathbb{N}} \subseteq \mathbb{N}$ satisfying

$$q_1\big(f_m - f_n\big) < 2^{-l}, \quad \text{for all } m,n \geqslant j_{1,l}.$$

Let $M_1 := \{j_{1,l} : l \in \mathbb{N}\}$. Since $\{f_n\}_{n\in M_1}$ is Cauchy, there exists an index $j_{2,1} \in M_1$ with $j_{2,1} > j_{1,1}$ such that $q_2(f_m - f_n) < 2^{-1}$, for all $m,n \geqslant j_{2,1}$ with $m,n \in M_1$. Also, $j_{2,1} = j_{1,p_1}$ for a certain $p_1 > 1$. As $\{f_n\}_{n\in M_1, n\geqslant j_{2,1}}$ is still Cauchy, we can choose an index $j_{2,2} > \max\{j_{2,1}, j_{1,2}\}$ such that $q_2(f_m - f_n) < 2^{-2}$, for all $m,n \geqslant j_{2,2}$ with $m,n \in M_1$. Also, we can choose $j_{2,2} = j_{1,p_2}$ for a certain $p_2 > p_1$. Assume that $j_{2,l-1}$ is already constructed for an arbitrary $l \in \mathbb{N}$. As $\{f_n\}_{n\in M_1, n\geqslant j_{2,l-1}}$ is still Cauchy, we can find an index $j_{2,l} > \max\{j_{2,l-1}, j_{1,l}\}$ such that $q_2(f_m - f_n) < 2^{-l}$, for all $m,n \geqslant j_{2,l}$ with $m,n \in M_1$. Also, we can choose $j_{2,l} = j_{1,p_l}$ for a certain $p_l > p_{l-1}$. Thereby we obtain a strictly increasing sequence $M_2 := \{j_{2,l}\}_{l\in\mathbb{N}} \subseteq M_1 := \{j_{1,l}\}_{l\in\mathbb{N}}$ such that

$$q_2\big(f_m - f_n\big) < 2^{-l}, \quad \text{for all } m,n \geqslant j_{2,l} \text{ with } m,n \in M_1,$$

holds for each $l \in \mathbb{N}$. Continue inductively to produce for each $k \in \mathbb{N}$ a strictly increasing sequence $M_k := \{j_{k,l}\}_{l\in\mathbb{N}} \subseteq \mathbb{N}$ such that $M_k \subseteq M_{k-1}$ and

$$q_k\big(f_m - f_n\big) < 2^{-l}, \quad \text{for all } m,n \geqslant j_{k,l} \text{ with } m,n \in M_{k-1}, \qquad (3.5)$$

hold for each $l \in \mathbb{N}$. Observe, for each $k \in \mathbb{N}$, that $\{f_{j_{k+1,l}}\}_{l\in\mathbb{N}}$ is a subsequence of $\{f_{j_{k,l}}\}_{l\in\mathbb{N}}$. For a fixed $k \in \mathbb{N}$ consider the telescoping sum

$$f_{j_{k,r}} - f_{j_{k,p}} = \sum_{l=p}^{r-1}\big(f_{j_{k,l+1}} - f_{j_{k,l}}\big), \quad \text{for } r > p \geqslant 1.$$

Due to the triangle inequality for q_k we derive, for all $r > p \geqslant 1$, that

$$
\begin{aligned}
q_k\big(f_{j_{k,r}} - f_{j_{k,p}}\big) &= q_k\left(\sum_{l=p}^{r-1}\big(f_{j_{k,l+1}} - f_{j_{k,l}}\big)\right) \\
&\leqslant \sum_{l=p}^{r-1} q_k\big(f_{j_{k,l+1}} - f_{j_{k,l}}\big) \\
&\overset{(3.5)}{<} \sum_{l=p}^{r-1}\frac{1}{2^l} = \frac{1}{2^p}\sum_{l=0}^{r-1-p}\frac{1}{2^l} \\
&= \frac{1}{2^p}\left(\frac{\left(\frac{1}{2}\right)^{r-1-p+1} - 1}{\frac{1}{2} - 1}\right)
\end{aligned}
$$

$$= \frac{1}{2^{p-1}} \underbrace{\left(1 - \left(\tfrac{1}{2}\right)^{r-p}\right)}_{<1,\,\text{as }r>p} < \frac{1}{2^{p-1}}.$$

Define for each $l \in \mathbb{N}$ a new sequence $\{g_l\}_{l\in\mathbb{N}}$ by $g_l := f_{j_{l,l}}$. Then $\{g_l\}_{l\in\mathbb{N}}$ is a subsequence of $\{f_n\}_{n\in\mathbb{N}}$. Fix $k \in \mathbb{N}$. For every $l > k$ and for certain $r > p > l$, the previous inequality shows that

$$q_k\left(g_{l+1} - g_l\right) = q_k\left(f_{j_{k,r}} - f_{j_{k,p}}\right) < \frac{1}{2^{p-1}} < \frac{1}{2^{l-1}}$$

holds. Accordingly,

$$\sum_{l=k+1}^{N} q_k\left(g_{l+1} - g_l\right) \leqslant \sum_{l=k+1}^{\infty} q_k\left(g_{l+1} - g_l\right) < \sum_{l=k+1}^{\infty} \frac{1}{2^{l-1}} = \sum_{l=k}^{\infty} \frac{1}{2^l} < \infty.$$

Since $k \in \mathbb{N}$ is arbitrary and $L_{\{q_k\}}$ has the (JRF)-property, Lemma 3.1.1 implies that

$$q_k\left(\sum_{l=1}^{\infty} |g_{l+1} - g_l|\right) \leqslant \sum_{l=1}^{\infty} q_k\left(g_{l+1} - g_l\right) < \infty, \quad \text{for all } k \in \mathbb{N}. \qquad (3.6)$$

Therefore the function defined by $g := \sum_{l=1}^{\infty} |g_{l+1} - g_l|$ is an element of $L_{\{q_k\}}$ and, by Lemma 2.3.1, surely satisfies $0 \leqslant g(w) < \infty$ for μ-almost every $w \in \Omega$.

Consider the set $E := \{w \in \Omega : g(w) = \infty\}$. Then E is a μ-null set and g is an absolutely convergent series pointwise on E^c; it follows that also $\sum_{l=1}^{\infty}(g_{l+1} - g_l)$ is pointwise convergent on E^c. Define on E^c the function $\tilde{f} := g_1 + \sum_{l=1}^{\infty}(g_{l+1} - g_l)$. Then $\tilde{f} - g_p = \sum_{l=p}^{\infty}(g_{l+1} - g_l)$, for each $p \in \mathbb{N}$. Defining $h_l := g_{l+1} - g_l$, for all $l \in \mathbb{N}$, the inequality

$$\left|\sum_{l=p}^{N} h_l\right| = |h_p + h_{p+1} + \ldots + h_N| \leqslant |h_p| + |h_{p+1}| + \ldots + |h_N| = \sum_{l=p}^{N} |h_l|$$

holds, for all $N \geqslant p$. For $p \in \mathbb{N}$ fixed, by taking the limit of both sides for $N \to \infty$ (pointwise on E^c), we obtain

$$\left|\sum_{l=p}^{\infty}(g_{l+1} - g_l)\right| = \left|\sum_{l=p}^{\infty} h_l\right| \leqslant \sum_{l=p}^{\infty} |h_l| = \sum_{l=p}^{\infty} |g_{l+1} - g_l| < \infty$$

which implies that (pointwise on E^c)

$$\left|\tilde{f} - g_p\right| \leqslant \sum_{l=p}^{\infty} |g_{l+1} - g_l| \overset{p\to\infty}{\longrightarrow} 0.$$

Thus, $\tilde{f}$ is the pointwise limit of $\{g_l\}_{l\in\mathbb{N}}$ on E^c. Additionally, as each q_k is a function semi-norm it follows from (3.6) that

$$
\begin{aligned}
q_k\big(\tilde{f}-g_p\big) &= q_k\left(\sum_{l=p}^{\infty}\big(g_{l+1}-g_l\big)\right)\\
&\leqslant q_k\left(\sum_{l=p}^{\infty}\big|g_{l+1}-g_l\big|\right)\\
&\leqslant \sum_{l=p}^{\infty} q_k\big(g_{l+1}-g_l\big) \overset{p\to\infty}{\longrightarrow} 0, \quad \text{for all } k\in\mathbb{N}.
\end{aligned}
$$

Accordingly, for fixed $k\in\mathbb{N}$, we have for all $n,p\in\mathbb{N}$ that

$$
\begin{aligned}
q_k\big(\tilde{f}-f_n\big) &= q_k\big(\tilde{f}-g_p+g_p-f_n\big)\\
&\leqslant q_k\big(\big|\tilde{f}-g_p\big|+\big|g_p-f_n\big|\big)\\
&\leqslant q_k\big(\tilde{f}-g_p\big)+q_k\big(g_p-f_n\big)\\
&= q_k\big(\tilde{f}-g_p\big)+q_k\big(f_{j_{p,p}}-f_n\big).
\end{aligned}
$$

Given $\varepsilon>0$ choose p such that $q_k\big(\tilde{f}-g_p\big)<\frac{\varepsilon}{2}$. Since $\{f_n\}_{n\in\mathbb{N}}$ is Cauchy, there is an index $N\geqslant j_{p,p}$ such that $q_k\big(f_n-f_m\big)<\frac{\varepsilon}{2}$, for all $m,n\geqslant N$. In particular, $q_k\big(f_{j_{p,p}}-f_n\big)<\frac{\varepsilon}{2}$, for all $n\geqslant N$. Hence, $q_k\big(\tilde{f}-f_n\big)<\varepsilon$, for all $n\geqslant N$. This shows that $\lim_{n\to\infty}q_k\big(\tilde{f}-f_n\big)=0$. Hence, $\tilde{f}\in L_{\{q_k\}}$ is the limit of the Cauchy sequence $\{f_n\}_{n\in\mathbb{N}}$ in the topology of $L_{\{q_k\}}$. This shows that $L_{\{q_k\}}$ is complete and, thus, is a Fréchet function space.

(i) $\Rightarrow$ (ii) Let $\{u_n\}_{n\in\mathbb{N}}\subseteq L_{\{q_k\}}^+$ be a sequence satisfying $\sum_{n=1}^{\infty}q_k(u_n)<\infty$, for all $k\in\mathbb{N}$. According to Remark 2.3.1 it suffices to show that $\sum_{n=1}^{\infty}u_n\in L_{\{q_k\}}^+$. Define for each $n\in\mathbb{N}$ the partial sum $s_n:=u_1+u_2+\ldots+u_n$. The sequence $\{s_n\}_{n\in\mathbb{N}}$ is Cauchy in $L_{\{q_k\}}$ as, for each $k\in\mathbb{N}$, we have whenever $m>n$ that

$$
q_k\big(s_m-s_n\big)=q_k\left(\sum_{l=n+1}^{m}u_l\right)\leqslant\sum_{l=n+1}^{m}q_k(u_l)\overset{m,n\to\infty}{\longrightarrow}0.
$$

As $L_{\{q_k\}}$ is complete the sequence $\{s_n\}_{n\in\mathbb{N}}$ converges to a function $f\in L_{\{q_k\}}$ in the topology of $L_{\{q_k\}}$.

In a first step we show that f is $\mathbb{R}$-valued μ-a.e.. As $f\in L_{\{q_k\}}$ we have $f=g+i\,h$, where g,h are $\mathbb{R}$-valued functions. Since all the s_n take their values in $[0,\infty]$, we can write $f-s_n=(g-s_n)+i\,h$, where h is the imaginary part of $f-s_n$, i.e., $h=\operatorname{Im}(f-s_n)$, for all $n\in\mathbb{N}$. Hence, $|h|\leqslant|f-s_n|$, for all $n\in\mathbb{N}$, and as each q_k is a function semi-norm we obtain

$$q_k(h) \leqslant q_k(f - s_n) \overset{n \to \infty}{\longrightarrow} 0, \quad \text{for all } k \in \mathbb{N}.$$

Accordingly, $h = 0$ μ-a.e. meaning that f is $\mathbb{R}$-valued μ-a.e..

In a second step we show that $f \geqslant 0$ μ-a.e.. As f is an $\mathbb{R}$-valued function we have $f = f^+ - f^-$ where $f^+ := \max\{f, 0\}$ and $f^- := \max\{-f, 0\}$. However, $s_n \geqslant 0$ by definition and therefore $f - s_n \leqslant f$, for all $n \in \mathbb{N}$. Consider, for a fixed n, the sets $A_n^+ := \{w \in \Omega : (f - s_n)(w) \geqslant 0\}$ and $A_n^- := \{w \in \Omega : (f - s_n)(w) \leqslant 0\}$. On A_n^+ the inequalities $f \geqslant s_n \geqslant 0$ imply that $f^- = 0$ and so

$$f \geqslant f - s_n \geqslant 0 = f^-.$$

Therefore $|f - s_n| \geqslant f^-$ is true on A_n^+. Note, that if $b \geqslant 0$ and $a \in \mathbb{R}$ satisfy $a \leqslant b$, then $|a - b| \geqslant a^-$. Hence, on A_n^- it follows that $|f - s_n| \geqslant f^-$ as well. So, the inequality $|f - s_n| \geqslant f^-$ holds μ-a.e. on Ω and is valid for all $n \in \mathbb{N}$. Again, we use the fact that each q_k is a function semi-norm and derive

$$q_k(f^-) \leqslant q_k(|f - s_n|) = q_k(f - s_n) \overset{n \to \infty}{\longrightarrow} 0, \quad \text{for all } k \in \mathbb{N}.$$

Accordingly, $f^- = 0$ μ-a.e. meaning that $f \geqslant 0$ μ-a.e.. Hence, $f \in L_{\{q_k\}}^+$.

In a last step we show that $f \geqslant s_n$, for all $n \in \mathbb{N}$. It is clear, for a fixed $l \in \mathbb{N}$, that $s_l \leqslant s_n$ holds for all $n \geqslant l$. Define again two sets $B_l^+ := \{w \in \Omega : f(w) \geqslant s_l(w)\}$ and $B_l^- := \{w \in \Omega : f(w) \leqslant s_l(w)\}$. Hence, on B_l^+, the function $s_l = \min\{s_l, f\}$ and therefore $s_l - \min\{s_l, f\} = 0$, implying that

$$\left|s_l - \min\{s_l, f\}\right| = 0 \leqslant |s_n - f| \quad \text{on } B_l^+, \text{ for all } n \geqslant l.$$

On B_l^-, $f = \min\{s_l, f\}$ and therefore $0 \leqslant s_l - f \leqslant s_n - f$, implying that

$$\left|s_l - \min\{s_l, f\}\right| = |s_l - f| \leqslant |s_n - f| \quad \text{on } B_l^-, \text{ for all } n \geqslant l.$$

Hence, $|s_l - \min\{s_l, f\}| \leqslant |s_n - f|$ holds μ-a.e. on Ω and is valid for all $n \geqslant l$. As each q_k is a function semi-norm it follows that

$$q_k(s_l - \min\{s_l, f\}) \leqslant q_k(s_n - f) \overset{n \to \infty}{\longrightarrow} 0, \quad \text{for all } k \in \mathbb{N}.$$

Accordingly, $s_l = \min\{s_l, f\}$ μ-a.e. for each $l \in \mathbb{N}$. As l was chosen arbitrarily, we can conclude that

$$\sum_{n=1}^{l} u_n = s_l \leqslant f, \quad \mu\text{-a.e., for all } l \in \mathbb{N}.$$

Taking the pointwise limit for $l \to \infty$ of the left side we obtain $\sum_{n=1}^{\infty} u_n \leqslant f$. Again, as each q_k is a function semi-norm and $f \in L_{\{q_k\}}$ we finally get

$$q_k\left(\sum_{n=1}^{\infty} u_n\right) \leqslant q_k(f) < \infty, \quad \text{for all } k \in \mathbb{N},$$

meaning that $\sum_{n=1}^{\infty} u_n \in L_{\{q_k\}}^+$. Thus, $L_{\{q_k\}}$ has the (JRF)-property. $\square$

From now on we will focus on complete, metrizable function spaces, i.e., Fréchet function spaces, over a σ-*finite measure space* (Ω, Σ, μ). The next lemma formulates a result which we obtained in the proof of Lemma 3.1.2. Its assertion turns out to be a useful tool for the forthcoming proofs and applications. For Banach function spaces, this result is well-known; see [26, Proposition 2.2] and the references given there.

Lemma 3.1.3

Let $X(\mu) = L_{\{q_k\}}$ be a Fréchet function space whose topology is generated by a fundamental, increasing sequence of function semi-norms $\{q_k\}_{k\in\mathbb{N}}$. Let $\{f_n\}_{n\in\mathbb{N}} \subseteq X(\mu)$ be a sequence which converges to f in the topology of $X(\mu)$. Then there exists a subsequence of $\{f_n\}_{n\in\mathbb{N}}$ which converges to f μ-a.e..

Proof:

Let $\{f_n\}_{n\in\mathbb{N}} \subseteq X(\mu)$ be a sequence that converges to an element $f \in X(\mu)$ in the topology of $X(\mu)$, that is,

$$\lim_{n\to\infty} q_k(f_n - f) = 0, \quad \text{for all } k \in \mathbb{N}.$$

Hence, $\{f_n\}_{n\in\mathbb{N}}$ is a Cauchy sequence in $X(\mu)$. In the proof of Lemma 3.1.2 it was shown that in this case there exists a subsequence $\{g_l\}_{l\in\mathbb{N}}$ of $\{f_n\}_{n\in\mathbb{N}}$ and a function $\tilde{f} \in X(\mu)$ which is on the one hand the μ-a.e. pointwise limit of the sequence $\{g_l\}_{l\in\mathbb{N}}$ and on the other hand the limit of $\{f_n\}_{n\in\mathbb{N}}$ in the topology of $X(\mu)$ meaning that

$$\lim_{n\to\infty} q_k(f_n - \tilde{f}) = 0, \quad \text{for all } k \in \mathbb{N}.$$

Using the fact that each q_k is a function semi-norm we obtain

$$q_k(f - \tilde{f}) = q_k(f - f_n + f_n - \tilde{f}) \;\leqslant\; q_k(|f - f_n| + |f_n - \tilde{f}|)$$
$$\leqslant\; q_k(f - f_n) + q_k(f_n - \tilde{f}) \xrightarrow{n\to\infty} 0,$$

for all $k \in \mathbb{N}$, and can therefore conclude that $\tilde{f} = f$. Thus, there exists a subsequence of $\{f_n\}_{n\in\mathbb{N}}$ (namely $\{g_l\}_{l\in\mathbb{N}}$) converging to f μ-a.e.. $\square$

Now we can show that $X(\mu)$ is continuously included in $M(\mu)$. For $X(\mu)$ a Banach function space, see [26, Proposition 2.2 (i)].

Proposition 3.1.1

Let $X(\mu) = L_{\{q_k\}}$ be a Fréchet function space whose topology is generated by a fundamental, increasing sequence of function semi-norms $\{q_k\}_{k\in\mathbb{N}}$. Furthermore, let the complete, metrizable space $M(\mu)$ be equipped with its topology of local convergence in measure. Then the natural inclusion map $i : X(\mu) \to M(\mu)$ is continuous.

Proof:

We apply the Closed Graph Theorem as stated in [18, p. 168]; see also the paragraph before Proposition 2.1.1. So, let $\{f_n\}_{n\in\mathbb{N}}$ be a sequence in $X(\mu) \subseteq M(\mu)$ which converges to $0 \in X(\mu)$ in the topology of $X(\mu)$ and such that $\{i(f_n)\}_{n\in\mathbb{N}}$ converges to a function $\tilde{f} \in M(\mu)$ in the topology of $M(\mu)$. We need to show that $\tilde{f} = 0$.

The fact that $\{f_n\}_{n\in\mathbb{N}}$ converges to 0 in the topology of $X(\mu)$ ensures, by Lemma 3.1.3, that there exists a subsequence $\{f_{n_m}\}_{m\in\mathbb{N}}$ of $\{f_n\}_{n\in\mathbb{N}}$ which converges to 0 μ-a.e.. It follows from Remark 2.2.2 (ii) that $\{f_{n_m}\}_{m\in\mathbb{N}}$ locally converges in measure to 0 as well. On the other hand, being a subsequence of $\{i(f_n)\}_{n\in\mathbb{N}}$ the sequence $\{i(f_{n_m})\}_{m\in\mathbb{N}} = \{f_{n_m}\}_{m\in\mathbb{N}}$ converges already to the function $\tilde{f} \in M(\mu)$ in the topology of $M(\mu)$ meaning that $\{f_{n_m}\}_{m\in\mathbb{N}}$ locally converges in measure to $\tilde{f}$. But then we can conclude by Remark 2.2.2 (i) that $\tilde{f} = 0$ in $M(\mu)$. $\square$

The continuity of the inclusion map implies the following result; see [26, Proposition 2.2] for Banach function spaces.

Corollary 3.1.1

Let $X(\mu) = L_{\{q_k\}}$ be a Fréchet function space whose topology is generated by a fundamental, increasing sequence of function semi-norms $\{q_k\}_{k\in\mathbb{N}}$. Then every Cauchy sequence in $X(\mu)$ admits a subsequence converging μ-a.e..

Proof:

Let $\{f_n\}_{n\in\mathbb{N}}$ be an arbitrary Cauchy sequence in $X(\mu)$. Then there exists a function $f \in X(\mu)$ such that $\{f_n\}_{n\in\mathbb{N}}$ converges to f in the topology of $X(\mu)$. By Proposition 3.1.1 it follows that $\{i(f_n)\}_{n\in\mathbb{N}}$ converges to $i(f)$ locally in measure with i being the identity map. By Remark 2.2.2 (iii) the sequence $\{f_n\}_{n\in\mathbb{N}}$ has a subsequence which converges to f μ-a.e.. $\square$

Now we can prove the second main result of this section. For $X(\mu)$, $Y(\mu)$ Banach function spaces, see [26, Lemma 2.7].

Proposition 3.1.2

Let $X(\mu)$ and $Y(\mu)$ be two Fréchet function spaces in $M(\mu)$ such that $X(\mu) \subseteq Y(\mu)$ as vector sublattices of $M(\mu)$. Then the natural inclusion map $j : X(\mu) \to Y(\mu)$ is continuous.

Proof:

We apply the Closed Graph Theorem 2.1.1. Let $\{f_n\}_{n\in\mathbb{N}}$ be a sequence in $X(\mu) \subseteq Y(\mu)$ such that $\{f_n\}_{n\in\mathbb{N}}$ converges to $0 \in X(\mu)$ in the topology of $X(\mu)$ and such that $\{j(f_n)\}_{n\in\mathbb{N}}$ converges to some function $\tilde{f} \in Y(\mu)$ in the topology of $Y(\mu)$ where $j : X(\mu) \to Y(\mu)$ is the natural inclusion map. We need to show that $\tilde{f} = 0$.

As $\{f_n\}_{n\in\mathbb{N}}$ converges to 0 in the topology of $X(\mu)$, Lemma 3.1.3 implies that there exists a subsequence $\{f_{n_m}\}_{m\in\mathbb{N}}$ of $\{f_n\}_{n\in\mathbb{N}}$ which converges to 0 μ-a.e.. The sequence $\{j(f_{n_m})\}_{m\in\mathbb{N}}$ in turn, being a subsequence of $\{j(f_n)\}_{n\in\mathbb{N}}$, converges to $\tilde{f}$ in the topology of $Y(\mu)$. Hence, Lemma 3.1.3 implies that there exists a subsequence $\{j(f_{n_{m_l}})\}_{l\in\mathbb{N}}$ of $\{j(f_{n_m})\}_{m\in\mathbb{N}}$ which converges to $\tilde{f}$ μ-a.e.. Now, since $\{j(f_{n_{m_l}})\}_{l\in\mathbb{N}} = \{f_{n_{m_l}}\}_{l\in\mathbb{N}}$ is a subsequence of $\{f_{n_m}\}_{m\in\mathbb{N}}$ the μ-a.e. limits have to be the same and we can conclude that $j(f) = 0 = \tilde{f}$. Thus, $j : X(\mu) \to Y(\mu)$ is continuous. $\qquad\square$

3.2 The vector measure m_T associated with T

Throughout this section let (Ω, Σ, μ) again be a σ-finite measure space. As usual, $X(\mu)$ will denote a Fréchet function space whose topology is generated by a fundamental, increasing sequence of function semi-norms $\{q_k\}_{k\in\mathbb{N}}$ whereas X will be a Fréchet space equipped with a fundamental, increasing sequence of semi-norms $\{p_k\}_{k\in\mathbb{N}}$. We will write $(X(\mu), \{q_k\}_{k\in\mathbb{N}})$ and $(X, \{p_k\}_{k\in\mathbb{N}})$ whenever we want to emphasize this. It will be *assumed throughout this section* that $X(\mu)$ contains all Σ-simple functions. Then $\chi_\Omega \in X(\mu)$ and it follows that also $L^\infty(\mu) \subseteq X(\mu)$. Furthermore, let $T : X(\mu) \to X$ be a continuous linear operator. By means of the operator T we define a finitely additive set function $m_T : \Sigma \to X$ by

$$m_T(A) := T(\chi_A), \quad \text{for } A \in \Sigma, \tag{3.7}$$

where χ_A is the characteristic function of A.

Recall that in Lemma 2.3.2 we have proven that $\mathrm{sim}(\Sigma)$ is dense in $X(\mu)$ whenever the Fréchet function space $X(\mu)$ contains the Σ-simple functions and has a σ-Lebesgue topology. Let us show that under the same conditions on $X(\mu)$ the finitely additive set function m_T becomes σ-additive.

Proposition 3.2.1

Let X be a Fréchet space, $X(\mu)$ be a Fréchet function space with σ-Lebesgue topology and $T : X(\mu) \to X$ be a continuous linear operator. Then m_T as defined in (3.7) is a vector measure.

Proof:

Let $\{A_j\}_{j \in \mathbb{N}} \subseteq \Sigma$ be any sequence of sets satisfying $A_j \downarrow_j \varnothing$, in which case $\{\chi_{A_j}\}_{j \in \mathbb{N}}$ is a sequence of functions satisfying $\chi_{A_j} \downarrow_j 0$ pointwise on Ω. Moreover, the inequality $|\chi_\Omega| \geqslant \chi_{A_j} \downarrow_j 0$ holds. The fact that $X(\mu)$ has a σ-Lebesgue topology implies that $\{\chi_{A_j}\}_{j \in \mathbb{N}}$ converges to 0 in the topology of $X(\mu)$; see Remark 2.3.2. By the continuity of T we obtain that $m_T(A_j) = T(\chi_{A_j})$ converges to 0 in the topology of X. Hence, m_T is σ-additive, i.e., a vector measure. $\quad\square$

We will refer to m_T as the *vector measure associated with T*. Of interest is the space of m_T-integrable functions $\mathcal{L}^1(m_T)$. The next result shows that $X(\mu)$ is always contained in it. For T taking values in a Banach space X, see [26, Proposition 4.4 (i)].

Proposition 3.2.2

Let $(X(\mu), \{q_k\}_{k \in \mathbb{N}})$ be a Fréchet function space with a σ-Lebesgue topology, X be a Fréchet space and m_T be the vector measure associated with a continuous linear operator $T : X(\mu) \to X$. Then each $f \in X(\mu)$ is m_T-integrable and $T(f\chi_A) = \int_A f \, dm_T$, for $A \in \Sigma$. In particular, $X(\mu) \subseteq \mathcal{L}^1(m_T)$.

Proof:

Let $f \in X(\mu)$. According to Lemma 2.3.2 there is a sequence of Σ-simple functions $\{s_n\}_{n \in \mathbb{N}} \subseteq \mathrm{sim}(\Sigma)$ which converges pointwise to f on Ω and which converges to f in the topology of $X(\mu)$. Fix $A \in \Sigma$ and consider the sequence $\{s_n \chi_A\}_{n \in \mathbb{N}} \subseteq X(\mu)$ as well as the function $f \chi_A \in X(\mu)$. It is clear that

$$\left| s_n \chi_A - f \chi_A \right| \leqslant \left| s_n - f \right|, \quad \text{for all } n \in \mathbb{N}.$$

This and the fact that each q_k is a function semi-norm imply that

$$q_k\left(s_n \chi_A - f \chi_A\right) \leqslant q_k\left(s_n - f\right) \overset{n \to \infty}{\longrightarrow} 0, \quad \text{for all } k \in \mathbb{N},$$

meaning that $\{s_n \chi_A\}_{n \in \mathbb{N}}$ converges to $f \chi_A$ in the topology of $X(\mu)$. Since T is continuous, the sequence $\{T(s_n \chi_A)\}_{n \in \mathbb{N}} \subseteq X$ converges to $T(f \chi_A) \in X$ in the

64

topology of X. Each function s_n, being Σ-simple, is of the form

$$s_n = \sum_{j=1}^{\ell(n)} \alpha_j^{(n)} \chi_{A_j^{(n)}}, \quad \text{for all } n \in \mathbb{N},$$

and so we can use the classical notation

$$\int_A s_n \, dm_T = \sum_{j=1}^{\ell(n)} \alpha_j^{(n)} m_T \left(A_j^{(n)} \cap A \right)$$

as stated in (2.27) and obtain, by the definition of m_T and the linearity of T, that

$$\int_A s_n \, dm_T = \sum_{j=1}^{\ell(n)} \alpha_j^{(n)} T \left(\chi_{A_j^{(n)} \cap A} \right) = T \left(\sum_{j=1}^{\ell(n)} \alpha_j^{(n)} \chi_{A_j^{(n)}} \chi_A \right) = T \left(s_n \chi_A \right),$$

for all $n \in \mathbb{N}$. Hence, $T(f\chi_A)$ is the limit of the sequence $\left\{ \int_A s_n \, dm_T \right\}_{n \in \mathbb{N}}$ in the topology of X, that is,

$$T(f\chi_A) = \lim_{n \to \infty} \int_A s_n \, dm_T.$$

Thus, there exists a sequence of Σ-simple functions $\{s_n\}_{n \in \mathbb{N}}$ which converges pointwise to f on Ω and such that $\left\{ \int_A s_n \, dm_T \right\}_{n \in \mathbb{N}}$ converges to the element $T(f\chi_A)$ in X. The set A was chosen arbitrarily and hence, this is true for all $A \in \Sigma$. Applying Proposition 2.4.1 we can conclude that f is m_T-integrable and $T(f\chi_A) = \int_A f \, dm_T$, for all $A \in \Sigma$. □

Concerning the null functions we have the following fact. For T being Banach-space-valued, see [26, Proposition 4.4 (ii)].

Lemma 3.2.1

Let $f \in \mathcal{L}^1(\mu)$ be a μ-null function. Then f is also an m_T-null function. In particular, $\mathcal{N}(\mu) \subseteq \mathcal{N}(m_T)$.

Proof:

Let $f \in \mathcal{L}^1(\mu)$ be an individual μ-null function, i.e., $f \in \mathcal{N}(\mu)$. Hence, $f = 0$ μ-a.e. on Ω and therefore also $f\chi_A = 0$ μ-a.e., for all $A \in \Sigma$. Then, for a fixed set $A \in \Sigma$,

$$\int_A f \, dm_T = T(f\chi_A) = T(0) = 0$$

and therefore also

$$p_k \left(\int_A f \, dm_T \right) = p_k \left(T(f\chi_A) \right) = p_k(0) = 0, \quad \text{for all } k \in \mathbb{N}.$$

As $A \in \Sigma$ was chosen arbitrarily this is true for all $A \in \Sigma$. We obtain that

$$p_k(m_T)(f) = \sup\left\{ p_k\left(\int_A f\, dm_T \right) : A \in \Sigma \right\} = 0, \quad \text{for all } k \in \mathbb{N}.$$

Thus, f is an m_T-null function, i.e., $f \in \mathcal{N}(m_T)$. $\qquad \square$

Denote by $L^1(m_T)$ the space of all classes of m_T-integrable functions (i.e., differing only on an m_T-null set). Namely,

$$L^1(m_T) := \mathcal{L}^1(m_T)/\mathcal{N}(m_T).$$

To show that $X(\mu)$ is included continuously into $L^1(m_T)$, consider the linear map $j_T : X(\mu) \to L^1(m_T)$ defined by $j_T(f) := f$. Observe, that this map is well-defined and not dependent on the representative f. To see this let $f, g \in \mathcal{M}(\mu)$ be two individual functions satisfying $f, g \in X(\mu)$ and differing only on a μ-null set. Hence, $f - g \in \mathcal{N}(\mu)$ and, because of Lemma 3.2.1, also $j_T(f - g) = f - g \in \mathcal{N}(m_T)$. Thus, f and g differ only on an m_T-null set and therefore determine the same element in $L^1(m_T)$. According to Proposition 3.2.2 we have

$$T(f) = T(f\chi_\Omega) = \int_\Omega f\, dm_T = \int_\Omega j_T(f)\, dm_T, \quad \text{for } f \in X(\mu). \tag{3.8}$$

For $X(\mu)$ a Banach function space and T a Banach-space-valued operator, the following fact occurs in [26, Proposition 4.4 (ii)].

Proposition 3.2.3

The linear map $j_T : X(\mu) \to L^1(m_T)$ is continuous.

Proof:

Let $f \in X(\mu)$ and fix a set $A \in \Sigma$. Then $|f\chi_A| \leqslant |f|$ and as each q_k is a function semi-norm, the inequality $q_k(f\chi_A) \leqslant q_k(f)$ holds, for all $k \in \mathbb{N}$. Fix $k \in \mathbb{N}$. The continuity of T implies that there exists an index $l_k \in \mathbb{N}$ and a constant $M_k > 0$ such that

$$p_k\big(T(f\chi_A)\big) \leqslant M_k\, q_{l_k}\big(f\chi_A\big) \leqslant M_k\, q_{l_k}(f), \quad \text{for } A \in \Sigma.$$

Keeping in mind that

$$\int_A j_T(f)\, dm_T = \int_\Omega j_T(f)\chi_A\, dm_T \overset{(3.8)}{=} T\big(f\chi_A\big),$$

66

we obtain that

$$p_k(m_T)(j_T(f)) = \sup\left\{ p_k\left(\int_A j_T(f)\,dm_T\right) : A \in \Sigma \right\} \leqslant M_k\, q_{l_k}(f).$$

Since $k \in \mathbb{N}$ is arbitrary, this shows that the linear map $j_T : X(\mu) \to L^1(m_T)$ is continuous. $\square$

The next result shows that $X(\mu)$ is continuously *embedded* in $L^1(m_T)$ whenever the m_T-null functions and the μ-null functions coincide.

Proposition 3.2.4

Whenever $\mathcal{N}(m_T) = \mathcal{N}(\mu)$, the continuous linear map $j_T : X(\mu) \to L^1(m_T)$ is injective. That is, $X(\mu)$ is continuously included in $L^1(m_T)$.

Proof:

Let $f \in X(\mu)$ satisfy $j_T(f) = 0$, that is, $j_T(f) \in L^1(m_T)$ is an m_T-null function. But, as $\mathcal{N}(m_T) \subseteq \mathcal{N}(\mu)$, it is also a μ-null function. Hence, $j_T(f) = f = 0$ μ-a.e.. Thus, j_T is injective. $\square$

We call a continuous linear operator $T : X(\mu) \to X$ *μ-determined* if the μ-null functions coincide with the m_T-null functions, i.e., $\mathcal{N}(\mu) = \mathcal{N}(m_T)$. In Proposition 3.2.4 we have seen that the μ-determinedness of the operator T causes the natural inclusion map j_T to be injective. For $X(\mu)$ a Banach function space and X a Banach space, see [26, Lemma 4.5].

Lemma 3.2.2

The following assertions for a continuous linear operator $T : X(\mu) \to X$ are equivalent:
  (i) T is μ-determined.
  (ii) $\mathcal{N}(\mu) = \mathcal{N}(m_T)$.
  (iii) $\mathcal{N}_0(\mu) = \mathcal{N}_0(m_T)$.

Proof:

(i) $\Leftrightarrow$ (ii) is clear by definition.

(i) $\Rightarrow$ (iii) Suppose that T is μ-determined, meaning that the m_T-null functions and the μ-null functions coincide. Let $A \in \mathcal{N}_0(\mu)$ be any μ-null set. Then $\chi_A \in \mathcal{N}(\mu)$ and by Lemma 3.2.1 $\chi_A \in \mathcal{N}(m_T)$ as well which means that the indefinite integral m_{T,χ_A} is the null vector measure. But this is equivalent to $A \in \mathcal{N}_0(m_T)$ since

$$0 = m_{T,\chi_A}(B) = \int_B \chi_A\,dm_T = m_T(B \cap A), \quad \text{for all } B \in \Sigma. \tag{3.9}$$

Thus, $\mathcal{N}_0(\mu) \subseteq \mathcal{N}_0(m_T)$.

Conversely, let $A \in \mathcal{N}_0(m_T)$ be any m_T-null set. By definition χ_A is then an m_T-null function. But, by the μ-determinedness of T, we have $\mathcal{N}(m_T) \subseteq \mathcal{N}(\mu)$ implying that χ_A is also a μ-null function. Hence, $A \in \mathcal{N}_0(\mu)$ and consequently $\mathcal{N}_0(m_T) \subseteq \mathcal{N}_0(\mu)$.

(iii) $\Rightarrow$ (i) Since $\mathcal{N}(\mu) \subseteq \mathcal{N}(m_T)$ is always true (see Lemma 3.2.1) it suffices to show that $\mathcal{N}(m_T) \subseteq \mathcal{N}(\mu)$ whenever the m_T-null sets and the μ-null sets coincide. Let $f \in \mathcal{N}(m_T)$ be any m_T-null function. By Remark 2.4.2 $f^{-1}(\mathbb{C}\backslash\{0\}) = \{w \in \Omega : f(w) \neq 0\}$ is then an m_T-null set. But, as $\mathcal{N}_0(m_T) = \mathcal{N}_0(\mu)$ holds, $f^{-1}(\mathbb{C}\backslash\{0\})$ is also a μ-null set. Thus, f is a μ-null function and $\mathcal{N}(m_T) \subseteq \mathcal{N}(\mu)$. $\square$

Recall that a σ-finite measure $\nu : \Sigma \to [0, \infty]$ is a *control measure* for m_T if the ν-null sets and the m_T-null sets coincide, i.e., if $\mathcal{N}_0(\nu) = \mathcal{N}_0(m_T)$. Remark 2.4.3 shows that ν can also be chosen as a finite measure. Lemma 3.2.2 asserts that μ is a control measure for m_T precisely when T is μ-determined.

Now we can show that the μ-determinedness of T is equivalent to the μ-determinedness of j_T; see also [26, Lemma 4.5 (ii)] for a special case.

Proposition 3.2.5

The operator $T : X(\mu) \to X$ is μ-determined if and only if the operator $j_T : X(\mu) \to L^1(m_T)$ is μ-determined.

Proof:

Assume that T is μ-determined, i.e., $\mathcal{N}(m_T) = \mathcal{N}(\mu)$. Consider the vector measure $m_{j_T} : \Sigma \to L^1(m_T)$ defined by $m_{j_T}(A) := j_T(\chi_A) = \chi_A$. Fix $A \in \mathcal{N}_0(m_{j_T})$. Then,

$$j_T(\chi_{A \cap B}) = m_{j_T}(B \cap A) = 0 \in L^1(m_T), \quad \text{for all } B \in \Sigma.$$

Since T is μ-determined, j_T is by Proposition 3.2.4 injective. Hence, $\chi_{B \cap A} = 0$ in $X(\mu)$ and therefore $m_T(B \cap A) = T(\chi_{B \cap A}) = T(0) = 0$, for all $B \in \Sigma$. Thus, A is also an m_T-null set and so $\mathcal{N}_0(m_{j_T}) \subseteq \mathcal{N}_0(m_T) = \mathcal{N}_0(\mu)$. It follows from Lemma 3.2.2 and Lemma 3.2.1, applied to $j_T : X(\mu) \to L^1(m_T)$, that $\mathcal{N}_0(m_{j_T}) = \mathcal{N}_0(\mu)$ meaning that j_T is μ-determined as well.

To show the converse direction assume that j_T is μ-determined and let $A \in \Sigma$ be m_T-null, i.e., $A \in \mathcal{N}_0(m_T)$. Then $\chi_A = 0$ in $L^1(m_T)$ meaning that $p_k(m_T)(\chi_A) = 0$, for all $k \in \mathbb{N}$. For each $B \subseteq A$, $\chi_B \leqslant \chi_A$ everywhere on Ω and thus,

$$p_k(m_T)(\chi_B) \leqslant p_k(m_T)(\chi_A) = 0, \quad \text{for all } k \in \mathbb{N}.$$

68

Therefore, $\chi_B = 0$ in $L^1(m_T)$. But $\chi_B = m_{j_T}(B)$, for all $B \subseteq A$, i.e., $m_{j_T}(B) = 0$ in $L^1(m_T)$, for all $B \subseteq A$. By the μ-determinedness of j_T we obtain $A \in \mathcal{N}_0(m_{j_T}) = \mathcal{N}_0(\mu)$. Hence, the m_T-null sets coincide with the μ-null sets and so T is μ-determined. $\square$

Another criterion for the μ-determinedness of the operator T is given in the following lemma. For a particular case see [26, Lemma 4.5 (iii)].

Proposition 3.2.6

Let $T : X(\mu) \to X$ be injective on the subset $\{\chi_A : A \in \Sigma\} \subseteq X(\mu)$. Then T is μ-determined.

Proof:

Suppose that T is injective on the subset $\{\chi_A : A \in \Sigma\}$ and let $B \in \mathcal{N}_0(m_T)$. Then $\chi_B \in X(\mu)$ and $T(\chi_B) = m_T(B) = 0$. As T is injective on $\{\chi_A : A \in \Sigma\}$, it follows that $\chi_B = 0$ in $X(\mu)$ and so $B \in \mathcal{N}_0(\mu)$. Thus, $\mathcal{N}_0(m_T) \subseteq \mathcal{N}_0(\mu)$. Lemma 3.2.1 and Lemma 3.2.2 then imply that T is μ-determined. $\square$

As the Σ-simple functions $\mathrm{sim}(\Sigma)$ are contained in $X(\mu)$, by assumption, and therefore also $\{\chi_A : A \in \Sigma\} \subseteq X(\mu)$ holds, we can conclude the following special case.

Corollary 3.2.1

Suppose that T is injective on $X(\mu)$. Then T is μ-determined. $\square$

3.3 The optimal domain and integral extension of the operator T

In the following result let (Ω, Σ, μ) be a σ-finite measure space, $X(\mu)$ be a Fréchet function space with a σ-Lebesgue topology and containing the Σ-simple functions, X be a Fréchet space, $T : X(\mu) \to X$ be a μ-determined, continuous linear operator and m_T be the vector measure associated with it. Let furthermore $j_T : X(\mu) \to L^1(m_T)$ be the continuous injection embedding $X(\mu)$ into $L^1(m_T)$; see Propositions 3.2.2, 3.2.3 and 3.2.4. Recall that $L^1(m_T)$ is a Fréchet function space over (Ω, Σ, μ) and has a σ-Lebesgue topology; see Section 2.4.

The following result shows that the *integration operator $I_{m_T} : L^1(m_T) \to X$*, defined by

$$I_{m_T}(f) := \int_\Omega f \, dm_T, \quad \text{for } f \in L^1(m_T),$$

is then a continuous, X-valued linear extension of T to the larger domain space $L^1(m_T)$, which is in a certain sense optimal. For the Banach function space setting we refer to Theorem 4.14 and Remark 4.15 of [26].

Theorem 3.3.1

The Fréchet function space $L^1(m_T)$ is the largest amongst all Fréchet function spaces over (Ω, Σ, μ) having a σ-Lebesgue topology into which $X(\mu)$ is continuously embedded and to which T admits an X-valued continuous linear extension. Moreover, such an extension of T is unique and is precisely the integration operator $I_{m_T} : L^1(m_T) \to X$.

Proof:

Let $Y(\mu)$ be any Fréchet function space over (Ω, Σ, μ) having a σ-Lebesgue topology such that $X(\mu) \subseteq Y(\mu)$ and such that T admits a continuous linear extension $\tilde{T} : Y(\mu) \to X$. According to Proposition 3.1.2 the natural embedding $j : X(\mu) \to Y(\mu)$ is necessarily continuous. We show that necessarily $Y(\mu) \subseteq L^1(m_T)$. As $\text{sim}(\Sigma) \subseteq X(\mu) \subseteq Y(\mu)$, we have $j(\chi_A) = \chi_A \in Y(\mu)$, for each $A \in \Sigma$. Therefore,

$$\tilde{T}(\chi_A) = \tilde{T}(j(\chi_A)) = \tilde{T}|_{X(\mu)}(\chi_A) = T(\chi_A), \quad \text{for all } A \in \Sigma.$$

By definition of the vector measures m_T and $m_{\tilde{T}}$ associated with T respectively $\tilde{T}$ this means that

$$m_{\tilde{T}}(A) = \tilde{T}(\chi_A) = T(\chi_A) = m_T(A), \quad \text{for all } A \in \Sigma.$$

Hence, the X-valued vector measures m_T and $m_{\tilde{T}}$ coincide. Since T is μ-determined we obtain $\mathcal{N}(m_{\tilde{T}}) = \mathcal{N}(m_T) = \mathcal{N}(\mu)$ meaning that $\tilde{T}$ is μ-determined as well. On the other hand, Proposition 3.2.2 applied to $\tilde{T}$ implies that $Y(\mu) \subseteq L^1(m_{\tilde{T}}) = L^1(m_T)$, meaning that $L^1(m_T)$ is "larger" than $Y(\mu)$. We still need to show that $I_{m_T} : L^1(m_T) \to X$ is a continuous linear extension of T from $X(\mu)$ to $L^1(m_T)$. Let $f \in X(\mu) \subseteq L^1(m_T)$. Then Proposition 3.2.2 yields

$$(I_{m_T} \circ j_T)(f) = I_{m_T}(j_T(f)) = I_{m_T}(f) = \int_\Omega f \, dm_T = T(f\chi_\Omega) = T(f).$$

Hence, I_{m_T} is indeed a continuous linear extension of T to $L^1(m_T)$.

The extension I_{m_T} is also unique. To see this let $\Lambda : L^1(m_T) \to X$ be another continuous linear extension of T, meaning that

$$I_{m_T}(f) = T(f) = \Lambda(f), \quad \text{for all } f \in X(\mu). \tag{3.10}$$

Since $L^1(m_T)$ contains the Σ-simple functions and has a σ-Lebesgue topology, Lemma

2.3.2 implies that $\text{sim}(\Sigma)$ is dense in $L^1(m_T)$. Now, let $f \in L^1(m_T)$ be arbitrarily chosen. Also by Lemma 2.3.2 there exists a sequence $\{s_n\}_{n \in \mathbb{N}} \subseteq \text{sim}(\Sigma)$ converging to f in the topology of $L^1(m_T)$. Then (3.10) yields that

$$I_{m_T}(f) = I_{m_T}\left(\lim_{n\to\infty} s_n\right) = \lim_{n\to\infty} I_{m_T}(s_n) = \lim_{n\to\infty} \Lambda(s_n) = \Lambda\left(\lim_{n\to\infty} s_n\right) = \Lambda(f).$$

Hence, $I_{m_T} = \Lambda$. $\square$

Chapter 4

Applications

In Chapter 4 we will apply the theory we have developed in Chapter 3 to take a closer look at some well-known operators $T : X(\mu) \to X$ defined on a Fréchet function space $X(\mu)$ and to study the vector measures m_T associated with them. The main object of interest, of course, will be the space $L^1(m_T)$ of m_T-integrable functions and one of the major problems will certainly be to decide whether or not $L^1(m_T)$ is strictly larger than $X(\mu)$.

The first operator we concentrate on in Section 4.1 will be the multiplication operator $M_g : X(\mu) \to X(\mu)$ defined by

$$M_g(f) := fg, \quad \text{for } f \in X(\mu),$$

where $g \in \mathcal{M} := \{g \in L^0(\mu) : g \cdot X(\mu) \subseteq X(\mu)\}$ is fixed. The vector measure $m_{M_g} : \Sigma \to X(\mu)$ associated with M_g is then given by

$$m_{M_g}(A) := M_g(\chi_A) = \chi_A g, \quad \text{for } A \in \Sigma.$$

In [26, Example 4.7] the respective investigations were made for the multiplication operator $M_g^r : L^r(\mu) \to L^r(\mu)$ $(1 \leqslant r < \infty)$, for $g \in L^\infty(\mu)$ fixed, defined on the Banach function space $L^r(\mu)$, with μ being a finite measure. There, the characterization of $L^1(m_{M_g^r})$ depended on whether g was "bounded away from 0" or not. It turned out that $L^1(m_{M_g^r}) = L^r(\mu)$ (in the first case) respectively $L^1(m_{M_g^r}) = \{\frac{1}{g} \cdot f : f \in L^r(\mu)\}$ (in the second case). In Section 4.1 we vary the situation by defining the multiplication operator on the Fréchet function space $L^{p-}([0,1])$ (Subsection 4.1.1) and on the Fréchet function space $L^p_{\text{loc}}(\mathbb{R})$ (Subsection 4.1.2) and investigate if similar observations can be made. In a further step we will also study the variation of m_{M_g}.

Section 4.2 deals with the Volterra operator $V_{p-} : L^{p-}([0,1]) \to L^{p-}([0,1])$ given by

$$V_{p-} : f \mapsto V_{p-}(f)(w) := \int_0^w f \, d\lambda, \quad \text{for } w \in [0,1],$$

where the resulting vector measure $m_{V_{p-}} : \mathcal{B}([0,1]) \to L^{p-}([0,1])$ associated with V_{p-} is given by

$$m_{V_{p-}} : A \mapsto m_{V_{p-}}(A) := V_{p-}(\chi_A)(t) = \int_A \chi_{[0,t]}(w) \, d\lambda(w), \quad \text{for } t \in [0,1].$$

The classical Volterra operator $V_p : L^p([0,1]) \to L^p([0,1])$ defined on the Banach function space $L^p([0,1])$, for $1 \leqslant p < \infty$, was the object of investigation in [28] and [26, Example 3.26 & Example 3.45]; the results again depended on p. For $p = 1$ the inclusion $L^1([0,1]) \subseteq L^1(m_{V_1})$ turned out to be proper with

$$L^1(m_{V_1}) = L^1(|m_{V_1}|) = L^1\big((1-t)\, d\lambda(t)\big).$$

For $1 < p < \infty$, however, all the inclusions

$$L^p([0,1]) \subseteq L^1(|m_{V_p}|) \subseteq L^1(m_{V_p})$$

were strict. In both cases the Bochner λ-integrability, respectively the Pettis λ-integrability, of $t \mapsto f\chi_{[t,1]}$ (where $f \in L^0([0,1])$ and $\chi_{[t,1]} \in L^p([0,1])$) played a major role in the investigations. So it is understandable that in Section 4.2 we search for similar results for V_{p-} being defined on the Fréchet function space $L^{p-}([0,1])$.

In the final Section 4.3 we will study the convolution operator $C_g^{p-} : L^{p-}(G) \to L^{p-}(G)$ defined by

$$C_g^{p-}(f) := f * g$$

where, for $g \in L^1(G)$ fixed,

$$(f * g)(x) := \int_G f(y)\, g(x - y) \, d\mu(y), \quad \text{for } \mu\text{-almost every } x \in G,$$

is the convolution of f and g on the compact Abelian group G. The vector measure $m_{C_g^{p-}} : \mathcal{B}(G) \to L^{p-}(G)$ associated with the convolution operator C_g^{p-} is then given by

$$m_{C_g^{p-}}(A) := C_g^{p-}(\chi_A) = \chi_A * g, \quad \text{for } A \in \mathcal{B}(G).$$

The respective investigations for the convolution operator $C_g^{(p)} : L^p(G) \to L^p(G)$, for $1 \leqslant p < \infty$, were done in [26, Chapter 7] and [25]. The main results state that the inclusion

$$L^p(G) \subseteq L^1\big(m_{C_g^{(p)}}\big)$$

is proper, for all non-zero functions $g \in L^1(G)\backslash L^p(G)$, whereas for $g \in L^p(G)$ we have the equalities

$$L^1\big(|m_{C_g^{(p)}}|\big) = L^1\big(m_{C_g^{(p)}}\big) = L^1(G).$$

There, the Bochner λ-integrability, this time of the function $F_g^{(p)} : G \to L^p(G)$ given by $F_g^{(p)}(x) = g(x - \cdot)$, played a central role in the course of the investigations. In Section 4.3 it will therefore be of interest to see whether the results differ a lot when C_g^{p-} is defined on $L^{p-}(G)$ rather than on $L^p(G)$.

4.1   Multiplication operators

4.1.1   Multiplication operators on $L^{p-}([0,1])$

Throughout this subsection we consider the finite measure space $([0,1], \mathcal{B}([0,1]), \lambda)$, where λ is Lebesgue measure and $\mathcal{B}([0,1])$ the σ-algebra of Borel measurable subsets of $[0,1]$. Let, for $p \in (1,\infty)$ fixed, $L^{p-}([0,1])$ be the Fréchet function space as defined in Example 2.3.1. Furthermore, denote by $L^0([0,1])$ the Lebesgue measurable functions $f : [0,1] \to \mathbb{C}$. Since $[0,1]$ is fixed, we will simply write L^0 and L^{p-} in place of $L^0([0,1])$ and $L^{p-}([0,1])$; no confusion will occur.

Consider the subset of L^0 given by

$$\begin{aligned} \mathcal{M}^{p-} &:= \mathcal{M}\big(L^{p-}, L^{p-}\big) \\ &:= \big\{g \in L^0 : gL^{p-} \subseteq L^{p-}\big\}, \end{aligned}$$

where $gL^{p-} := \{gh : h \in L^{p-}\}$ and $g \in \mathcal{M}^{p-}$ is fixed. Associated with g define the *multiplication operator* $M_g^{p-} : L^{p-} \to L^{p-}$ by

$$M_g^{p-}(f) := fg, \quad \text{for } f \in L^{p-}.$$

It is clear that M_g^{p-} is a linear operator. Moreover, as $\chi_{[0,1]} \in L^{p-}$, it follows that necessarily $g \in L^{p-}$.

Proposition 4.1.1
For each $g \in \mathcal{M}^{p-}$ and $p \in (1,\infty)$, the multiplication operator $M_g^{p-} : L^{p-} \to L^{p-}$ is continuous.

Proof:

We apply the Closed Graph Theorem 2.1.1. Let $\{f_n\}_{n\in\mathbb{N}} \subseteq L^{p-}$ be any sequence which converges to 0 in the topology of L^{p-} and such that $\{M_g^{p-}(f_n)\}_{n\in\mathbb{N}}$ converges to a function $f_0 \in L^{p-}$ in the topology of L^{p-}. We need to show that $f_0 = 0$.

Since $\{M_g^{p-}(f_n)\}_{n\in\mathbb{N}}$ converges to f_0 in the topology of L^{p-}, Lemma 3.1.3 implies that there exists a subsequence $\{M_g^{p-}(f_{n_m})\}_{m\in\mathbb{N}}$ of $\{M_g^{p-}(f_n)\}_{n\in\mathbb{N}}$ converging to f_0 λ-a.e. on $[0,1]$. On the other hand, $\{f_{n_m}\}_{m\in\mathbb{N}}$ is a subsequence of $\{f_n\}_{n\in\mathbb{N}}$ and therefore converges to 0 in the topology of L^{p-}. Applying Lemma 3.1.3 again we find that there exists a subsequence $\{f_{n_{m_l}}\}_{l\in\mathbb{N}}$ of $\{f_{n_m}\}_{m\in\mathbb{N}}$ which converges to 0 λ-a.e. on $[0,1]$. Multiplying the functions $f_{n_{m_l}}$ with g we obtain, for λ-almost every $w \in [0,1]$,

$$\lim_{l\to\infty} M_g^{p-}\left(f_{n_{m_l}}\right)(w) = \lim_{l\to\infty} f_{n_{m_l}}(w)g(w) = 0 \cdot g(w) = 0.$$

However, $\{M_g^{p-}(f_{n_m})\}_{m\in\mathbb{N}}$ converges pointwise to f_0 λ-a.e. and, being a subsequence of $\{M_g^{p-}(f_{n_m})\}_{m\in\mathbb{N}}$, the same is true for $\{M_g^{p-}\left(f_{n_{m_l}}\right)\}_{l\in\mathbb{N}}$. Hence, $f_0 = 0$ and we can conclude that M_g^{p-} is continuous. $\square$

In the following pages we are going to study the vector measure $m_{M_g^{p-}}$ associated with M_g^{p-}, i.e., the vector measure $m_{M_g^{p-}} : \mathcal{B}([0,1]) \to L^{p-}$ defined by

$$m_{M_g^{p-}}(A) := M_g^{p-}(\chi_A) = \chi_A g, \quad \text{for } A \in \mathcal{B}([0,1]). \tag{4.1}$$

Since L^{p-} contains the $\mathcal{B}([0,1])$-simple functions and has a σ-Lebesgue topology, $m_{M_g^{p-}}$ is (by Proposition 3.2.1) indeed a vector measure. Recall that each $\varphi \in \left(L^{p-}\right)^*$ induces the scalar measure $\langle m_{M_g^{p-}}, \varphi \rangle : \mathcal{B}([0,1]) \to \mathbb{C}$ given by

$$\langle m_{M_g^{p-}}, \varphi \rangle(A) := \langle m_{M_g^{p-}}(A), \varphi \rangle, \quad \text{for } A \in \mathcal{B}([0,1]),$$

where in our case this measure becomes (see (i) of Example 2.3.1)

$$\langle m_{M_g^{p-}}(A), \varphi \rangle = \int_0^1 M_g^{p-}(\chi_A)\varphi \, d\lambda \stackrel{(4.1)}{=} \int_0^1 \chi_A g\varphi \, d\lambda = \int_A g\varphi \, d\lambda, \tag{4.2}$$

for all $A \in \mathcal{B}([0,1])$. We are mainly interested in the space of $m_{M_g^{p-}}$-integrable functions. Recall that a function $f : [0,1] \to \mathbb{C}$ is $m_{M_g^{p-}}$-integrable if it is integrable with respect to each scalar measure $\langle m_{M_g^{p-}}, \varphi \rangle$, for $\varphi \in \left(L^{p-}\right)^*$, and if, for each $A \in \mathcal{B}([0,1])$, there exists an element $\int_A f \, dm_{M_g^{p-}} \in L^{p-}$ satisfying

$$\left\langle \int_A f \, dm_{M_g^{p-}}, \varphi \right\rangle = \int_A f \, d\langle m_{M_g^{p-}}, \varphi \rangle,$$

for all $\varphi \in \left(L^{p-}\right)^*$. It follows, for each $A \in \mathcal{B}([0,1])$, that

$$\int_A f \, d\langle m_{M_g^{p-}}, \varphi \rangle \overset{(4.2)}{=} \int_A fg\varphi \, d\lambda = \int_0^1 f\chi_A g\varphi \, d\lambda = \langle f\chi_A g, \varphi \rangle.$$

Accordingly, since φ is arbitrary, the indefinite integral of $f \in L^1(m_{M_g^{p-}})$ over $A \in \mathcal{B}([0,1])$ is given by

$$\int_A f \, dm_{M_g^{p-}} = f\chi_A g \tag{4.3}$$

and, hence, the set function $m_{M_g^{p-},f} : \mathcal{B}([0,1]) \to L^{p-}$ associated with the indefinite integral of f is defined by

$$m_{M_g^{p-},f}(A) := \int_A f \, dm_{M_g^{p-}} = f\chi_A g, \quad \text{for } A \in \mathcal{B}([0,1]).$$

According to the Orlicz-Pettis Theorem 2.1.3 $m_{M_g^{p-},f}$ is again a vector measure.

To apply the theory of Chapter 3 we need to know when M_g^{p-} is λ-determined.

Proposition 4.1.2

Let $1 < p < \infty$ and $g \in \mathcal{M}^{p-}$. The operator $M_g^{p-} : L^{p-} \to L^{p-}$ is λ-determined if and only if $g(w) \neq 0$ for λ-almost every $w \in [0,1]$.

Proof:

Suppose that $g \neq 0$ λ-a.e. on $[0,1]$ fails to hold. Then $\lambda\left(g^{-1}(\{0\})\right) > 0$ and so $B := g^{-1}(\{0\}) \notin \mathcal{N}_0(\lambda)$. On the other hand, for any Borel set $A \subseteq B$ we have

$$m_{M_g^{p-}}(A) = M_g^{p-}(\chi_A) = g\chi_A = 0 \in L^{p-}.$$

Hence, $B \in \mathcal{N}_0(m_{M_g^{p-}})$. So, $\mathcal{N}_0(\lambda) \neq \mathcal{N}_0(m_{M_g^{p-}})$. The contrapositive statement yields that $\mathcal{N}_0(\lambda) = \mathcal{N}_0(m_{M_g^{p-}})$, i.e., M_g^{p-} is λ-determined, implies that $g \neq 0$, λ-a.e. on $[0,1]$.

Let now $g \neq 0$ λ-a.e. on $[0,1]$. Choose any set $A \in \mathcal{N}_0(m_{M_g^{p-}})$. Then, in particular, $m_{M_g^{p-}}(A) = \chi_A g = 0$ in L^{p-}. But, since $g \neq 0$ λ-a.e. on $[0,1]$, this means that $\chi_A = 0$ λ-a.e. on $[0,1]$, i.e., $\lambda(A) = 0$. Thus, $A \in \mathcal{N}_0(\lambda)$. Keeping in mind that $\mathcal{N}_0(\lambda) \subseteq \mathcal{N}_0(m_{M_g^{p-}})$ is always true (by Lemma 3.2.1) we can conclude that $\mathcal{N}_0(\lambda) = \mathcal{N}_0(m_{M_g^{p-}})$. Therefore, M_g^{p-} is λ-determined. $\quad\square$

From now on we will always assume M_g^{p-} to be λ-determined. Since L^{p-} has a σ-Lebesgue topology and contains the $\mathcal{B}([0,1])$-simple functions we know from Proposition 3.2.2 that each $f \in L^{p-}$ is $m_{M_g^{p-}}$-integrable, i.e., that

$$L^{p-} \subseteq L^1(m_{M_g^{p-}}),$$

and, for each $f \in L^{p-}$, that the equations

$$M_g^{p-}(f\chi_A) = \int_A f\,dm_{M_g^{p-}} \overset{(4.3)}{=} f\chi_A g, \quad \text{for } A \in \mathcal{B}([0,1]),$$

hold. Moreover, it follows from the λ-determinedness of M_g^{p-} that

$$\mathcal{N}(\lambda) = \mathcal{N}(m_{M_g^{p-}}) \quad \text{resp.} \quad \mathcal{N}_0(\lambda) = \mathcal{N}_0(m_{M_g^{p-}})$$

and, from Theorem 3.3.1, that $L^1(m_{M_g^{p-}})$ is the optimal domain of the operator M_g^{p-} and the optimal extension of M_g^{p-} is the integration operator $I_{m_{M_g^{p-}}} : L^1(m_{M_g^{p-}}) \rightarrow L^{p-}$ given by

$$I_{m_{M_g^{p-}}}(f) := \int_0^1 f\,dm_{M_g^{p-}} = fg, \quad \text{for } f \in L^1(m_{M_g^{p-}}).$$

Furthermore, since the space L^{p-} is reflexive we know from Section 2.4 that

$$L^1(m_{M_g^{p-}}) = L_w^1(m_{M_g^{p-}}).$$

But, we still do not know whether $L^1(m_{M_g^{p-}})$ is strictly larger than L^{p-} and whether it is possible to characterize the space $L^1(m_{M_g^{p-}})$. This is what we intend to investigate in the following part.

First of all, let us give a characterization of $L^1(m_{M_g^{p-}})$.

Proposition 4.1.3

Let $g \in \mathcal{M}^{p-}$ satisfy $g \neq 0$ λ-a.e. on $[0,1]$. Then,

$$L^1(m_{M_g^{p-}}) = \{f \in L^0 : fg \in L^{p-}\}.$$

Proof:

Since $g \in \mathcal{M}^{p-}$ and $\mathrm{sim}(\mathcal{B}([0,1])) \subseteq L^{p-}$ we obtain that $sg \in L^{p-}$, for all $s \in \mathrm{sim}(\mathcal{B}([0,1]))$. In particular, choosing $s = \chi_{[0,1]}$ shows that necessarily $g \in L^{p-}$.

Now, let $f \in L^1(m_{M_g^{p-}})$. According to Proposition 2.4.1 there exists a sequence $\{s_n\}_{n \in \mathbb{N}} \subseteq \mathrm{sim}(\mathcal{B}([0,1]))$ such that $\{s_n\}_{n \in \mathbb{N}}$ converges pointwise to f and $\{\int_0^1 s_n\,dm_{M_g^{p-}}\}_{n \in \mathbb{N}}$ converges to an element $\int_0^1 f\,dm_{M_g^{p-}} \in L^{p-}$ in the topology of

L^{p-}. Lemma 3.1.3, on the other hand, ensures that $\left\{\int_0^1 s_n \, dm_{M_g^{p-}}\right\}_{n\in\mathbb{N}}$ admits a subsequence $\left\{\int_0^1 s_{n_m} \, dm_{M_g^{p-}}\right\}_{m\in\mathbb{N}}$ converging λ-a.e. on $[0,1]$ to $\int_0^1 f \, dm_{M_g^{p-}}$. But, since $\{s_n\}_{n\in\mathbb{N}}$ converges to f pointwise on $[0,1]$, we have

$$
\lim_{m\to\infty}\left(\int_0^1 s_{n_m}\, dm_{M_g^{p-}}\right)(w) \overset{(4.3)}{=} \lim_{m\to\infty}\left(s_{n_m}\chi_{[0,1]}g\right)(w)
$$
$$
= \lim_{m\to\infty}\left(s_{n_m}g\right)(w)
$$
$$
= (fg)(w),
$$

for λ-almost every $w \in [0,1]$. Hence, $fg = \int_0^1 f \, dm_{M_g^{p-}} \in L^{p-}$. This establishes one inclusion.

Conversely, let $f \in L^0$ satisfy $fg \in L^{p-}$. Since $f \in L^0$, we can choose a sequence $\{s_n\}_{n\in\mathbb{N}} \subseteq \mathrm{sim}\left(\mathcal{B}([0,1])\right)$ such that $|s_n| \leqslant |f|$, for all $n \in \mathbb{N}$, and $\{s_n\}_{n\in\mathbb{N}}$ converges pointwise to f on $[0,1]$. Since $g \in \mathcal{M}^{p-}$ and $\mathrm{sim}\left(\mathcal{B}([0,1])\right) \subseteq L^{p-}$ it follows that $s_n g \in L^{p-}$, for all $n \in \mathbb{N}$. On the other hand, $fg \in L^{p-}$ and consequently $f\chi_A g \in L^{p-}$, for all $A \in \mathcal{B}([0,1])$. Moreover, $|s_n g| \leqslant |fg|$, for all $n \in \mathbb{N}$, and the sequence $\{s_n g\}_{n\in\mathbb{N}}$ converges to fg pointwise on $[0,1]$. The σ-Lebesgue topology of L^{p-} guarantees that the sequence $\{s_n g\}_{n\in\mathbb{N}}$ converges to fg also in the topology of L^{p-}. Each q_k being a continuous function semi-norm (see Example 2.3.1) we obtain, for any $A \in \mathcal{B}([0,1])$ and $n \in \mathbb{N}$, that

$$
0 \leqslant q_k\left(\int_A s_n \, dm_{M_g^{p-}} - f\chi_A g\right) \overset{(4.3)}{=} q_k\left(s_n\chi_A g - f\chi_A g\right) \leqslant q_k\left(s_n g - fg\right),
$$

for all $k \in \mathbb{N}$. Since

$$
\lim_{n\to\infty} q_k\left(s_n g - fg\right) = 0, \quad \text{for all } k \in \mathbb{N},
$$

it follows, for each $A \in \mathcal{B}([0,1])$, that the sequence $\left\{\int_A s_n \, dm_{M_g^{p-}}\right\}_{n\in\mathbb{N}}$ converges to $f\chi_A g$ in the topology of L^{p-}. Proposition 2.4.1 now implies that $f \in L^1(m_{M_g^{p-}})$ and that, moreover, $\int_A f \, dm_{M_g^{p-}} = f\chi_A g$, for all $A \in \mathcal{B}([0,1])$. $\qquad\square$

Let us now investigate whether $L^1(m_{M_g^{p-}})$ is strictly larger than L^{p-} or, in other words, whether there exists a function $f \in L^1(m_{M_g^{p-}})$ that is not an element of L^{p-}. By using the characterization of $L^1(m_{M_g^{p-}})$ given in Proposition 4.1.3 this would mean that $f \notin L^{r_k}$, for at least one $k \in \mathbb{N}$, although $fg \in L^{p-}$.

Recall, for $g \in \mathcal{M}^{p-}$, that M_g^{p-} is assumed to be λ-determined and hence, $g \neq 0$

λ-a.e. on $[0,1]$. Therefore it is possible to write

$$f = \tfrac{1}{g} \cdot fg, \quad \lambda\text{-a.e. on } [0,1].$$

It seems to be the case that the answer to our question depends on the properties of g and, moreover, on the properties of $\frac{1}{g}$. Whenever $\frac{1}{g} \in L^\infty$ it is clear that $\frac{1}{g} fg = f \in L^{p-}$. So, the question is: What happens if $\frac{1}{g} \notin L^\infty$?

To continue our investigations we need the definition of a special type of operator-valued measure. For a Fréchet space X let $L(X)$ be the space of all continuous linear operators of X into itself. Then $L_s(X)$ is defined to be the space $L(X)$ equipped with the topology of pointwise convergence on X, that is, the topology of uniform convergence on all finite subsets of X. A σ-additive measure $P : \Sigma \to L_s(X)$ is said to be a *spectral measure* if it satisfies the following two conditions:

(i) $P(A \cap B) = P(A)P(B)$, for all $A, B \in \Sigma$.
(ii) $P(\Omega) = \mathrm{id}$.

Here, id is the identity operator in X.

For the following discussion we make use of a notable connection between the vector measure $m_{M_g^{p-}}$ and the spectral measure $\tilde{P} : \mathcal{B}([0,1]) \to L_s(L^{p-})$ given by

$$\tilde{P}(A) : f \mapsto f\chi_A, \quad \text{for } f \in L^{p-}.$$

The spectral measure $\tilde{P}$ was investigated in [1]. There, the following notation was used. Define, for $v \in L^0$ fixed, the vector space

$$D_p(\tilde{M}_v^{p-}) := \left\{ h \in L^{p-} : hv \in L^{p-} \right\} \subseteq L^{p-}.$$

Then $D_p(\tilde{M}_v^{p-})$ is the maximal domain of the linear operator $\tilde{M}_v^{p-} : D_p(\tilde{M}_v^{p-}) \to L^{p-}$ defined by $h \to hv$, for $h \in D_p(\tilde{M}_v^{p-})$. In [1, Proposition 18] it was established, for $v \in L^0$, that

$$D_p(\tilde{M}_v^{p-}) = L^{p-} \quad \text{if and only if} \quad v \in L^1(\tilde{P}) = \bigcap_{1 \leqslant s < \infty} L^s. \tag{4.4}$$

Note that $D_p(\tilde{M}_v^{p-}) = L^{p-}$ corresponds precisely to $v \in \mathcal{M}^{p-}$ with $\tilde{M}_v^{p-} = M_v^{p-}$, that is,

$$\mathcal{M}^{p-} = \bigcap_{1 \leqslant s < \infty} L^s. \tag{4.5}$$

Since, for $g \in \mathcal{M}^{p-} \subseteq L^{p-}$, it is the case that

$$\tilde{P}(A)(g) = g\chi_A = m_{M_g^{p-}}(A), \quad \text{for } A \in \mathcal{B}([0,1]),$$

we can use this result for identifying a set of functions g for which the spaces L^{p-} and $L^1(m_{M_g^{p-}})$ coincide. We point out that $\mathcal{M}^{p-}$ is actually independent of p.

Proposition 4.1.4

Let $g \in \mathcal{M}^{p-} = \bigcap_{1 \leqslant s < \infty} L^s$. Then,

$$L^1(m_{M_g^{p-}}) = L^{p-} \quad \text{if and only if} \quad \frac{1}{g} \in \mathcal{M}^{p-} = \bigcap_{1 \leqslant s < \infty} L^s.$$

Proof:

Suppose that $\frac{1}{g} \in \mathcal{M}^{p-} = \bigcap_{1 \leqslant s < \infty} L^s$. Fix $f \in L^1(m_{M_g^{p-}})$. Since $g \in \mathcal{M}^{p-}$, it follows from Proposition 4.1.3 that $fg \in L^{p-}$. Since $\frac{1}{g} \in \mathcal{M}^{p-}$, it follows that $f = \frac{1}{g} fg = M_{1/g}^{p-}(fg) \in L^{p-}$. So, the inclusion $L^1(m_{M_g^{p-}}) \subseteq L^{p-}$ holds. On the other hand, the inclusion $L^{p-} \subseteq L^1(m_{M_g^{p-}})$ follows from Proposition 3.2.2. Thus, $L^1(m_{M_g^{p-}}) = L^{p-}$.

Conversely, suppose that $L^1(m_{M_g^{p-}}) = L^{p-}$. Since $g \in \mathcal{M}^{p-}$, we know by (4.4) and (4.5) that $D_p(\tilde{M}_g^{p-}) = L^{p-}$. Choose an arbitrary function $f \in L^{p-}$. As M_g^{p-} is λ-determined, $g \neq 0$ λ-a.e. on $[0,1]$ and we can write $f = \frac{1}{g} fg$ where $\frac{1}{g} f \in L^0$ satisfies $\frac{1}{g} fg \in L^{p-}$. By the characterization of $L^1(m_{M_g^{p-}})$ in Proposition 4.1.3 we can conclude that $\frac{1}{g} f \in L^1(m_{M_g^{p-}})$ and hence, by assumption, also $\frac{1}{g} f \in L^{p-}$. Therefore, $f \in \{h \in L^{p-} : \frac{1}{g} h \in L^{p-}\} = D_p(\tilde{M}_{1/g}^{p-})$. But, f was chosen arbitrarily, and so $L^{p-} \subseteq D_p(\tilde{M}_{1/g}^{p-})$. On the other hand, $D_p(\tilde{M}_{1/g}^{p-}) \subseteq L^{p-}$ always holds. Thus, $D_p(\tilde{M}_{1/g}^{p-}) = L^{p-}$ which is, according to (4.4), equivalent to $\frac{1}{g} \in \bigcap_{1 \leqslant s < \infty} L^s$. $\square$

For every $1 < p < \infty$, the function $g(w) = w$, for $w \in [0,1]$, satisfies $g \in \mathcal{M}^{p-} = \bigcap_{1 \leqslant s < \infty} L^s$, but $\frac{1}{g} \notin \mathcal{M}^{p-}$. In particular, this together with Proposition 4.1.4 shows that the containment $L^{p-} \subseteq L^1(m_{M_g^{p-}})$ is proper.

A function $g \in L^0$ such that both $g, \frac{1}{g} \in (\bigcap_{1 \leqslant s < \infty} L^s) \backslash L^\infty$ is exhibited in the following example.

Example 4.1.1

Let $\{F_l\}_{l=0}^\infty \subseteq \mathcal{B}([0,1])|_{[0,\frac{1}{2}]}$ be the partition

$$F_0 := \left[\frac{1}{6}, \frac{1}{2}\right], \text{ and } F_l := \left[\frac{1}{2} - \sum_{j=1}^{l+1} \left(\frac{1}{3}\right)^j, \frac{1}{2} - \sum_{j=1}^{l} \left(\frac{1}{3}\right)^j\right), \text{ for } l \in \mathbb{N},$$

of the interval $\left[0, \frac{1}{2}\right]$. Since

$$\sum_{j=1}^{l} \left(\tfrac{1}{3}\right)^j = \frac{1 - \left(\tfrac{1}{3}\right)^{l+1}}{1 - \tfrac{1}{3}} - 1 = \tfrac{3}{2}\left(\tfrac{1}{3} - \tfrac{1}{3^{l+1}}\right) = \tfrac{1}{2}\left(1 - \tfrac{1}{3^l}\right) \tag{4.6}$$

and consequently

$$\tfrac{1}{2} - \sum_{j=1}^{l} \left(\tfrac{1}{3}\right)^j \overset{(4.6)}{=} \tfrac{1}{2} - \tfrac{1}{2}\left(1 - \tfrac{1}{3^l}\right) = \tfrac{1}{2 \cdot 3^l}$$

the intervals F_l can also be written as

$$F_0 = \left[\tfrac{1}{2 \cdot 3^1}, \tfrac{1}{2 \cdot 3^0}\right], \text{ and } F_l = \left[\tfrac{1}{2 \cdot 3^{l+1}}, \tfrac{1}{2 \cdot 3^l}\right), \text{ for } l \in \mathbb{N}.$$

Hence,

$$\lambda(F_l) = \tfrac{1}{2 \cdot 3^l} - \tfrac{1}{2 \cdot 3^{l+1}} = \tfrac{1}{2} \cdot \tfrac{2}{3^{l+1}} = \tfrac{1}{3^{l+1}}, \tag{4.7}$$

for all $l \in \mathbb{N}_0$. Moreover, let $\{E_m\}_{m=0}^{\infty} \subseteq \mathcal{B}([0,1])|_{(\frac{1}{2},1]}$ be the partition

$$E_0 := \left(\tfrac{1}{2}, \tfrac{5}{6}\right], \text{ and } E_m := \left(\tfrac{1}{2} + \sum_{j=1}^{m}\left(\tfrac{1}{3}\right)^j, \tfrac{1}{2} + \sum_{j=1}^{m+1}\left(\tfrac{1}{3}\right)^j\right], \text{ for } m \in \mathbb{N},$$

of the interval $\left(\tfrac{1}{2}, 1\right]$. Again the intervals E_m can be rewritten. Since

$$\tfrac{1}{2} + \sum_{j=1}^{m}\left(\tfrac{1}{3}\right)^j \overset{(4.6)}{=} \tfrac{1}{2} + \tfrac{1}{2}\left(1 - \tfrac{1}{3^m}\right) = 1 - \tfrac{1}{2 \cdot 3^m}$$

we obtain, for $m \in \mathbb{N}_0$,

$$E_m = \left(1 - \tfrac{1}{2 \cdot 3^m}, 1 - \tfrac{1}{2 \cdot 3^{m+1}}\right].$$

Then,

$$\lambda(E_m) = 1 - \tfrac{1}{2 \cdot 3^{m+1}} - 1 + \tfrac{1}{2 \cdot 3^m} = \tfrac{1}{3^{m+1}}, \tag{4.8}$$

for all $m \in \mathbb{N}_0$. Define a function $g : [0,1] \to \mathbb{C}$ by

$$\begin{aligned}
g(w) &:= \sum_{l=0}^{\infty}(l+1)\chi_{F_l}(w) + \sum_{m=0}^{\infty}\tfrac{1}{m+1}\chi_{E_m}(w) \\
&= \sum_{l=1}^{\infty}l\chi_{F_{l-1}}(w) + \sum_{m=1}^{\infty}\tfrac{1}{m}\chi_{E_{m-1}}(w).
\end{aligned}$$

Then g is obviously measurable, bounded on $\left[\tfrac{1}{2}, 1\right]$ but, unbounded on $\left[0, \tfrac{1}{2}\right]$. More-

over,
$$g(w) \leqslant \sum_{l=1}^{\infty} l\chi_{F_{l-1}}(w) + \sum_{m=1}^{\infty} m\chi_{E_{m-1}}(w),$$

for all $w \in [0,1]$. Hence, for each $1 \leqslant s < \infty$, we obtain that

$$
\begin{aligned}
q_s(g) \quad &:= \quad \left(\int_0^1 \left| \sum_{l=1}^{\infty} l\chi_{F_{l-1}} + \sum_{m=1}^{\infty} \frac{1}{m}\chi_{E_{m-1}} \right|^s d\lambda \right)^{1/s} \\
&\leqslant \quad \left(\int_0^1 \left(\sum_{l=1}^{\infty} l\chi_{F_{l-1}} + \sum_{m=1}^{\infty} m\chi_{E_{m-1}} \right)^s d\lambda \right)^{1/s} \\
&= \quad \left(\sum_{l=1}^{\infty} l^s \lambda(F_{l-1}) + \sum_{m=1}^{\infty} m^s \lambda(E_{m-1}) \right)^{1/s} \\
&\overset{(4.7),(4.8)}{=} \quad \left(\sum_{l=1}^{\infty} l^s \left(\tfrac{1}{3^l} \right) + \sum_{m=1}^{\infty} m^s \left(\tfrac{1}{3^m} \right) \right)^{1/s} \\
&= \quad 2^{1/s} \left(\sum_{n=1}^{\infty} \frac{n^s}{3^n} \right)^{1/s} < \infty.
\end{aligned}
$$

Accordingly, $g \in \bigcap_{1 \leqslant s < \infty} L^s$. On the other hand, $\frac{1}{g} : [0,1] \to \mathbb{C}$ is of the form

$$
\begin{aligned}
\tfrac{1}{g}(w) \quad &= \quad \sum_{l=0}^{\infty} \tfrac{1}{l+1}\chi_{F_l}(w) + \sum_{m=0}^{\infty} (m+1)\chi_{E_m}(w) \\
&= \quad \sum_{l=1}^{\infty} \tfrac{1}{l}\chi_{F_{l-1}}(w) + \sum_{m=1}^{\infty} m\chi_{E_{m-1}}(w).
\end{aligned}
$$

Note that $\frac{1}{g}$ is bounded on $\left[0, \frac{1}{2}\right]$, unbounded on $\left(\frac{1}{2}, 1\right]$ and satisfies

$$\tfrac{1}{g}(w) \leqslant \sum_{l=1}^{\infty} l\chi_{F_{l-1}}(w) + \sum_{m=1}^{\infty} m\chi_{E_{m-1}}(w),$$

for all $w \in [0,1]$. In analogy to g, for each $1 \leqslant s < \infty$, we obtain that

$$
\begin{aligned}
q_s\left(\tfrac{1}{g}\right) \quad &:= \quad \left(\int_0^1 \left| \sum_{l=1}^{\infty} \tfrac{1}{l}\chi_{F_{l-1}} + \sum_{m=1}^{\infty} m\chi_{E_{m-1}} \right|^s d\lambda \right)^{1/s} \\
&\leqslant \quad 2^{1/s} \left(\sum_{n=1}^{\infty} \frac{n^s}{3^n} \right)^{1/s} < \infty.
\end{aligned}
$$

So, both $g, \frac{1}{g} \in \bigcap_{1 \leqslant s < \infty} L^s$ and we can conclude by Proposition 4.1.4 that $L^1(m_{M_g^{p-}}) = L^{p-}$, for each $1 < p < \infty$. ◄

Until now we have not thought about the variation of $m_{M_g^{p-}}$. Fix $p \in (1, \infty)$ and $g \in \mathcal{M}^{p-}$. For each $k \in \mathbb{N}$, let $(m_{M_g^{p-}})_k : \mathcal{B}([0,1]) \to L^{r_k}$ be the local-Banach-space-valued vector measure (2.28) given by

$$(m_{M_g^{p-}})_k(A) := \chi_A g, \quad \text{for } A \in \mathcal{B}([0,1]).$$

For $k \in \mathbb{N}$ fixed, the variation of $(m_{M_g^{p-}})_k$ is calculated via

$$
\begin{aligned}
|(m_{M_g^{p-}})_k|(A) &= \sup_\pi \sum_{j=1}^{l} \|(m_{M_g^{p-}})_k(A_j)\|_{r_k} \\
&= \sup_\pi \sum_{j=1}^{l} \|\chi_{A_j} g\|_{r_k} \\
&= \sup_\pi \sum_{j=1}^{l} \left(\int_0^1 |\chi_{A_j} g|^{r_k} \, d\lambda \right)^{1/r_k} \\
&= \sup_\pi \sum_{j=1}^{l} \left(\int_{A_j} |g|^{r_k} \, d\lambda \right)^{1/r_k},
\end{aligned}
\tag{4.9}
$$

for all $A \in \mathcal{B}([0,1])$, where $\pi = \{A_j\}_{j=1}^{l}$ is any finite partition of A.

Note that the variation of $(m_{M_g^{p-}})_k$ needs not to be finite. To see this, let $g := \chi_{[0,1]}$ and choose, for $l \in \mathbb{N}$ fixed, the partition

$$A_j := \left[\tfrac{j-1}{l}, \tfrac{j}{l}\right), \text{ for } j = 1, \ldots, l-1, \text{ and } A_l := \left[\tfrac{l-1}{l}, 1\right]$$

of $[0,1]$. Then, for each $j \in \{1, \ldots, l\}$, we obtain

$$\sum_{j=1}^{l} \left(\int_{A_j} |g|^{r_k} \, d\lambda \right)^{1/r_k} = \sum_{j=1}^{l} (\lambda(A_j))^{1/r_k} = \sum_{j=1}^{l} \left(\tfrac{1}{l}\right)^{1/r_k} = l \cdot \left(\tfrac{1}{l}\right)^{1/r_k} = l^{1-(1/r_k)} = l^{1/s_k},$$

where s_k is the conjugate exponent of r_k and, thus, satisfies $\frac{1}{s_k} > 0$. So, if $l \to \infty$ it follows that $l^{1/s_k} \to \infty$. Hence, for $g = \chi_{[0,1]}$ the variation of $(m_{M_g^{p-}})_k$ and thus, the variation of $m_{M_g^{p-}}$ is infinite.

It seems that the variation of $m_{M_g^{p-}}$ could depend on the function g. Actually, the next result shows that this is not so.

Proposition 4.1.5

Let $g \in \mathcal{M}^{p-} \setminus \{0\}$. Then the variation of $m_{M_g^{p-}}$ is infinite.

Proof:

Let $g \in \mathcal{M}^{p^-}\setminus\{0\} = \left(\bigcap_{1 \leqslant s < \infty} L^s\right)\setminus\{0\}$. Then $g \in L^{p^-}\setminus\{0\}$ (see also page 75) and consequently $|g| \neq 0$ in L^{p^-}. On the other hand, $|g| \in L^{p^-}$ means that $|g|^{r_k} \in L^1$, for all $k \in \mathbb{N}$. So we can find $A \in \mathcal{B}([0,1])$ and $k \in \mathbb{N}$ such that

$$0 < \int_A |g|^{r_k}\, d\lambda < \infty.$$

Fix such an $A \in \mathcal{B}([0,1])$ and $k \in \mathbb{N}$ and let

$$\int_A |g|^{r_k}\, d\lambda =: \alpha.$$

Define a set function $\nu : \mathcal{B}(A \cap [0,1]) \to [0, \infty)$ by

$$\nu(B) := \int_B |g|^{r_k}\, d\lambda, \quad \text{for all } B \in \mathcal{B}(A \cap [0,1])$$

which is a finite, positive measure on $\mathcal{B}(A \cap [0,1])$. Since the Lebesgue measure is non-atomic on the real line, [9, p. 26], the measure ν is non-atomic on $\mathcal{B}(A \cap [0,1])$. Pick an arbitrary $l \in \mathbb{N}$. According to Lemma 2.2.2 there exists a partition $\{A_j\}_{j=1}^l \subseteq \mathcal{B}([0,1])$ of A satisfying

$$\nu(A_j) = \int_{A_j} |g|^{r_k}\, d\lambda = \tfrac{\alpha}{l}, \quad \text{for all } j = 1, \ldots, l.$$

Thus, we have

$$\sum_{j=1}^l \left(\int_{A_j} |g|^{r_k}\, d\lambda \right)^{1/r_k} = \sum_{j=1}^l \left(\nu(A_j)\right)^{1/r_k}$$

$$= \sum_{j=1}^l \left(\tfrac{\alpha}{l}\right)^{1/r_k} = l \cdot \left(\tfrac{\alpha}{l}\right)^{1/r_k} = l^{1-(1/r_k)} \cdot \alpha^{1/r_k} = l^{1/s_k} \cdot \alpha^{1/r_k},$$

where s_k is the conjugate exponent of r_k. Letting $l \to \infty$ we obtain that $l^{1/s_k} \cdot \alpha^{1/r_k} \to \infty$ which shows that the variation of $m_{M_g^{p^-}}$ is infinite. $\square$

4.1.2 Multiplication operators on $L_{\text{loc}}^p(\mathbb{R})$

In this subsection we will investigate the multiplication operator again, this time, however, defined on $L_{\text{loc}}^p(\mathbb{R})$. Consider the σ-finite measure space $(\mathbb{R}, \mathcal{B}(\mathbb{R}), \lambda)$, where λ is Lebesgue measure and $\mathcal{B}(\mathbb{R})$ is the σ-algebra of Lebesgue measurable subsets of $\mathbb{R}$. Let, for $p \in (1, \infty)$ fixed, $L_{\text{loc}}^p(\mathbb{R})$ be the Fréchet function space as defined in Example 2.3.2. Furthermore, denote by $L^0(\mathbb{R})$ the Lebesgue measurable functions

$f : \mathbb{R} \to \mathbb{C}$.

Define by $\mathcal{M}_{\text{loc}}^p$ the subset of measurable functions

$$\mathcal{M}_{\text{loc}}^p \;:=\; \mathcal{M}\big(L_{\text{loc}}^p(\mathbb{R}), L_{\text{loc}}^p(\mathbb{R})\big)$$
$$:=\; \big\{ g \in L^0(\mathbb{R}) : g\, L_{\text{loc}}^p(\mathbb{R}) \subseteq L_{\text{loc}}^p(\mathbb{R}) \big\},$$

where $g\, L_{\text{loc}}^p(\mathbb{R}) := \{ gh : h \in L_{\text{loc}}^p(\mathbb{R}) \}$. Fix $g \in \mathcal{M}_{\text{loc}}^p$. Associate with g the *multiplication operator* $M_{g,\text{loc}}^p : L_{\text{loc}}^p(\mathbb{R}) \to L_{\text{loc}}^p(\mathbb{R})$ defined by

$$M_{g,\text{loc}}^p(f) := fg, \quad \text{for } f \in L_{\text{loc}}^p(\mathbb{R}).$$

Clearly, $M_{g,\text{loc}}^p$ is a linear operator. It is continuous as well. The proof of the continuity of $M_{g,\text{loc}}^p$ follows the lines of the proof of Proposition 4.1.1 as we now show. Observe that it follows from $\chi_\mathbb{R} \in L_{\text{loc}}^p(\mathbb{R})$ that $g \in L_{\text{loc}}^p(\mathbb{R})$.

Proposition 4.1.6

For each $p \in (1, \infty)$, the multiplication operator $M_{g,loc}^p : L_{loc}^p(\mathbb{R}) \to L_{loc}^p(\mathbb{R})$ is continuous.

Proof:

We use the Closed Graph Theorem 2.1.1 again. Let $\{f_n\}_{n \in \mathbb{N}} \subseteq L_{\text{loc}}^p(\mathbb{R})$ be any sequence which converges to 0 in the topology of $L_{\text{loc}}^p(\mathbb{R})$ and such that $\big\{ M_{g,\text{loc}}^p(f_n) \big\}_{n \in \mathbb{N}}$ converges to a function $f_0 \in L_{\text{loc}}^p(\mathbb{R})$ in the topology of $L_{\text{loc}}^p(\mathbb{R})$. We need to show that $f_0 = 0$.

Since $\big\{ M_{g,\text{loc}}^p(f_n) \big\}_{n \in \mathbb{N}}$ converges to f_0 in the topology of $L_{\text{loc}}^p(\mathbb{R})$ we know by Lemma 3.1.3 that there exists a subsequence $\big\{ M_{g,\text{loc}}^p(f_{n_m}) \big\}_{m \in \mathbb{N}}$ of $\big\{ M_{g,\text{loc}}^p(f_n) \big\}_{n \in \mathbb{N}}$ converging to f_0 λ-a.e. on $\mathbb{R}$. But, being a subsequence of $\{f_n\}_{n \in \mathbb{N}}$, it follows that $\{f_{n_m}\}_{m \in \mathbb{N}}$ converges to 0 in the topology of $L_{\text{loc}}^p(\mathbb{R})$. Applying Lemma 3.1.3 again we obtain a subsequence $\big\{ f_{n_{m_l}} \big\}_{l \in \mathbb{N}}$ of $\{f_{n_m}\}_{m \in \mathbb{N}}$ which converges to 0 λ-a.e. on $\mathbb{R}$. By multiplying the functions $f_{n_{m_l}}$ with g we have, for λ-almost every $w \in \mathbb{R}$,

$$\lim_{l \to \infty} M_{g,\text{loc}}^p\big(f_{n_{m_l}}\big)(w) = \lim_{l \to \infty} f_{n_{m_l}}(w) g(w) = 0 \cdot g(w) = 0.$$

But $\big\{ M_{g,\text{loc}}^p(f_{n_m}) \big\}_{m \in \mathbb{N}}$ converges already to f_0 λ-a.e. on $\mathbb{R}$, so the same has to hold for the subsequence $\big\{ M_{g,\text{loc}}^p(f_{n_{m_l}}) \big\}_{l \in \mathbb{N}}$. Thus, $f_0 = 0$ and it follows that $M_{g,\text{loc}}^p$ is continuous. $\square$

The aim here is to study the vector measure $m_{M_{g,\text{loc}}^p}$ associated with the multiplication operator $M_{g,\text{loc}}^p$, i.e., the vector measure $m_{M_{g,\text{loc}}^p} : \mathcal{B}(\mathbb{R}) \to L_{\text{loc}}^p(\mathbb{R})$ defined by

$$m_{M_{g,\mathrm{loc}}^p}(A) := M_{g,\mathrm{loc}}^p(\chi_A) = \chi_A g, \quad \text{for } A \in \mathcal{B}(\mathbb{R}). \tag{4.10}$$

Note that $m_{M_{g,\mathrm{loc}}^p}$ is, by Proposition 3.2.1, indeed a vector measure as $L_{\mathrm{loc}}^p(\mathbb{R})$ contains the $\mathcal{B}(\mathbb{R})$-simple functions and has a σ-Lebesgue topology (see Example 2.3.2 (iv)). Observe, for each $\varphi \in \left(L_{\mathrm{loc}}^p(\mathbb{R})\right)^*$, that the scalar measure $\langle m_{M_{g,\mathrm{loc}}^p}, \varphi \rangle :$ $\mathcal{B}(\mathbb{R}) \to \mathbb{C}$ is given by

$$\langle m_{M_{g,\mathrm{loc}}^p}, \varphi \rangle(A) := \langle m_{M_{g,\mathrm{loc}}^p}(A), \varphi \rangle, \quad \text{for } A \in \mathcal{B}(\mathbb{R}),$$

which can also be expressed as follows:

$$\langle m_{M_{g,\mathrm{loc}}^p}(A), \varphi \rangle \overset{(4.10)}{=} \int_{\mathbb{R}} M_{g,\mathrm{loc}}^p(\chi_A)\varphi \, d\lambda \overset{(4.10)}{=} \int_{\mathbb{R}} \chi_A g \varphi \, d\lambda = \int_A g\varphi \, d\lambda, \tag{4.11}$$

for each $A \in \mathcal{B}(\mathbb{R})$. Note, since $g \in L_{\mathrm{loc}}^p(\mathbb{R})$, that $g\varphi \in L^1(\mathbb{R})$ for each $\varphi \in \left(L_{\mathrm{loc}}^p(\mathbb{R})\right)^*$; see Example 2.3.2 (i). Again, a measurable function $f : \mathbb{R} \to \mathbb{C}$ is $m_{M_{g,\mathrm{loc}}^p}$-integrable if it is integrable with respect to each scalar measure $\langle m_{M_{g,\mathrm{loc}}^p}, \varphi \rangle$, for $\varphi \in \left(L_{\mathrm{loc}}^p(\mathbb{R})\right)^*$, and if, for each $A \in \mathcal{B}(\mathbb{R})$, there exists an element $\int_A f \, dm_{M_{g,\mathrm{loc}}^p} \in L_{\mathrm{loc}}^p(\mathbb{R})$ satisfying

$$\left\langle \int_A f \, dm_{M_{g,\mathrm{loc}}^p}, \varphi \right\rangle = \int_A f \, d\langle m_{M_{g,\mathrm{loc}}^p}, \varphi \rangle,$$

for all $\varphi \in \left(L_{\mathrm{loc}}^p(\mathbb{R})\right)^*$. Because of (4.11) we obtain for each $A \in \mathcal{B}(\mathbb{R})$ that

$$\int_A f \, d\langle m_{M_{g,\mathrm{loc}}^p}, \varphi \rangle = \int_A f g \varphi \, d\lambda = \int_{\mathbb{R}} f \chi_A g \varphi \, d\lambda, \quad \text{for } \varphi \in \left(L_{\mathrm{loc}}^p(\mathbb{R})\right)^*.$$

Accordingly, $fg \in L_{\mathrm{loc}}^p(\mathbb{R})$ and so the indefinite integral of $f \in L^1\left(m_{M_{g,\mathrm{loc}}^p}\right)$ is given by

$$\int_A f \, dm_{M_{g,\mathrm{loc}}^p} = f \chi_A g \in L_{\mathrm{loc}}^p(\mathbb{R}), \quad \text{for } A \in \mathcal{B}(\mathbb{R}). \tag{4.12}$$

Indeed, to see this observe (via (4.11)) that

$$\left| \langle m_{M_{g,\mathrm{loc}}^p}, \varphi \rangle \right|(A) = \int_A |g\varphi| \, d\lambda,$$

for $A \in \mathcal{B}(\mathbb{R})$ and $\varphi \in \left(L_{\mathrm{loc}}^p(\mathbb{R})\right)^*$, with $g\varphi \in L^1(\mathbb{R})$ because $g \in \mathcal{M}_{\mathrm{loc}}^p \subseteq L_{\mathrm{loc}}^p(\mathbb{R})$. So, if f is $m_{M_{g,\mathrm{loc}}^p}$-integrable, then

$$\int_{\mathbb{R}} |f| \, |g| \, |\varphi| \, d\lambda = \int_{\mathbb{R}} |f| \, d\left|\langle m_{M_{g,\mathrm{loc}}^p}, \varphi \rangle\right| < \infty, \quad \text{for } \varphi \in \left(L_{\mathrm{loc}}^p(\mathbb{R})\right)^*.$$

Since $L^p_{loc}(\mathbb{R})$ is reflexive, this implies that $fg \in L^p_{loc}(\mathbb{R})$. The set function $m_{M^p_{g,loc},f}$: $\mathcal{B}(\mathbb{R}) \to L^p_{loc}(\mathbb{R})$ associated with the indefinite integral is given by

$$m_{M^p_{g,loc},f}(A) := \int_A f \, dm_{M^p_{g,loc}} = f\chi_A g.$$

By the Orlicz-Pettis Theorem 2.1.3 it is again a vector measure. Repeating the arguments in the proof of Proposition 4.1.2 one can show that $M^p_{g,loc}$ is λ-determined if and only if $g \neq 0$ λ-a.e. on $\mathbb{R}$ – a property which from now on we will assume that $M^p_{g,loc}$ always satisfies.

Let us see, whether $L^1(m_{M^p_{g,loc}})$ admits a similar characterization as for the multiplication operator M^{p-}_g.

Proposition 4.1.7

Let $p \in (1, \infty)$ and $g \in \mathcal{M}^p_{loc}$ satisfy $g \neq 0$ λ-a.e. on $\mathbb{R}$. Then,

$$L^1(m_{M^p_{g,loc}}) = \{f \in L^0(\mathbb{R}) : fg \in L^p_{loc}(\mathbb{R})\}.$$

Proof:

The fact that $g \in \mathcal{M}^p_{loc} \subseteq L^p_{loc}(\mathbb{R})$ and $\mathrm{sim}(\mathcal{B}(\mathbb{R})) \subseteq L^p_{loc}(\mathbb{R})$ ensures that $sg \in L^p_{loc}(\mathbb{R})$, for all $s \in \mathrm{sim}(\mathcal{B}(\mathbb{R}))$.

Let $f \in L^1(m_{M^p_{g,loc}})$. Due to Proposition 2.4.1 there exists a sequence $\{s_n\}_{n\in\mathbb{N}} \subseteq \mathrm{sim}(\mathcal{B}(\mathbb{R}))$ such that $\{s_n\}_{n\in\mathbb{N}}$ converges pointwise to f and $\{\int_\mathbb{R} s_n \, dm_{M^p_{g,loc}}\}_{n\in\mathbb{N}}$ converges to an element $\int_\mathbb{R} f \, dm_{M^p_{g,loc}}$ in the topology of $L^p_{loc}(\mathbb{R})$. According to Lemma 3.1.3, $\{\int_\mathbb{R} s_n \, dm_{M^p_{g,loc}}\}_{n\in\mathbb{N}}$ admits a subsequence $\{\int_\mathbb{R} s_{n_m} \, dm_{M^p_{g,loc}}\}_{m\in\mathbb{N}}$ converging to $\int_\mathbb{R} f \, dm_{M^p_{g,loc}}$ λ-a.e. on $\mathbb{R}$. However, the sequence $\{s_n\}_{n\in\mathbb{N}}$ and thus, also $\{s_{n_m}\}_{m\in\mathbb{N}}$, converges to f pointwise on $\mathbb{R}$ and so we obtain

$$\lim_{m\to\infty} \left(\int_\mathbb{R} s_{n_m} \, dm_{M^p_{g,loc}}\right)(w) \overset{(4.12)}{=} \lim_{m\to\infty} (s_{n_m}\chi_\mathbb{R} g)(w) = \lim_{m\to\infty} (s_{n_m} g)(w) = (fg)(w),$$

for λ-almost every $w \in \mathbb{R}$. Hence, $fg = \int_\mathbb{R} f \, dm_{M^p_{g,loc}} \in L^p_{loc}(\mathbb{R})$ which establishes one inclusion.

Conversely, let $f \in L^0(\mathbb{R})$ satisfy $fg \in L^p_{loc}(\mathbb{R})$. The fact that $f \in L^0(\mathbb{R})$ guarantees that we can find a sequence $\{s_n\}_{n\in\mathbb{N}} \subseteq \mathrm{sim}(\mathcal{B}(\mathbb{R}))$ such that $|s_n| \leqslant |f|$, for all $n \in \mathbb{N}$, and $\{s_n\}_{n\in\mathbb{N}}$ converges pointwise to f on $\mathbb{R}$. Since $g \in L^p_{loc}(\mathbb{R})$ and $\mathrm{sim}(\mathcal{B}(\mathbb{R})) \subseteq L^p_{loc}(\mathbb{R})$ we obtain that $s_n g \in L^p_{loc}(\mathbb{R})$, for all $n \in \mathbb{N}$. Moreover, $fg \in L^p_{loc}(\mathbb{R})$ and as a consequence also $f\chi_A g \in L^p_{loc}(\mathbb{R})$, for all $A \in \mathcal{B}(\mathbb{R})$. In addition, $|s_n g| \leqslant |fg|$, for all $n \in \mathbb{N}$, and the sequence $\{s_n g\}_{n\in\mathbb{N}}$ converges to fg

pointwise on $\mathbb{R}$. Due to the σ-Lebesgue topology of $L^p_{\text{loc}}(\mathbb{R})$ the sequence $\{s_n g\}_{n \in \mathbb{N}}$ converges to fg in the topology of $L^p_{\text{loc}}(\mathbb{R})$ as well. Since each q_k is a continuous function semi-norm in $L^p_{\text{loc}}(\mathbb{R})$ (see Example 2.3.2) we obtain, for $A \in \mathcal{B}(\mathbb{R})$ and $n \in \mathbb{N}$, that

$$0 \leqslant q_k \left(\int_A s_n \, dm_{M^p_{g,\text{loc}}} - f\chi_A g \right) \overset{(4.12)}{=} q_k\big(s_n \chi_A g - f\chi_A g\big) \leqslant q_k\big(s_n g - fg\big),$$

for all $k \in \mathbb{N}$. But,

$$\lim_{n \to \infty} q_k\big(s_n g - fg\big) = 0, \quad \text{for all } k \in \mathbb{N},$$

and so we can conclude, for each $A \in \mathcal{B}(\mathbb{R})$, that the sequence $\big\{\int_A s_n \, dm_{M^p_{g,\text{loc}}}\big\}_{n \in \mathbb{N}}$ converges to $f\chi_A g$ in the topology of $L^p_{\text{loc}}(\mathbb{R})$. It follows from Proposition 2.4.1 that $f \in L^1\big(m_{M^p_{g,\text{loc}}}\big)$ and, moreover, that $\int_A f \, dm_{M^p_{g,\text{loc}}} = f\chi_A g$, for all $A \in \mathcal{B}(\mathbb{R})$. $\square$

Let us write down here the properties of $m_{M^p_{g,\text{loc}}}$ and $L^1\big(m_{M^p_{g,\text{loc}}}\big)$ that follow from Chapter 3. First of all, since $L^p_{\text{loc}}(\mathbb{R})$ has a σ-Lebesgue topology and contains the $\mathcal{B}(\mathbb{R})$-simple functions, Proposition 3.2.2 implies that each $f \in L^p_{\text{loc}}(\mathbb{R})$ is $m_{M^p_{g,\text{loc}}}$-integrable, i.e., that

$$L^p_{\text{loc}}(\mathbb{R}) \subseteq L^1\big(m_{M^p_{g,\text{loc}}}\big)$$

and, for each $f \in L^p_{\text{loc}}(\mathbb{R})$, that the equation

$$M^p_{g,\text{loc}}(f\chi_A) = \int_A f \, dm_{M^p_{g,\text{loc}}} \overset{(4.12)}{=} f\chi_A g, \quad \text{for } A \in \mathcal{B}(\mathbb{R}),$$

holds. On the other hand, since we are assuming the λ-determinedness of $M^p_{g,\text{loc}}$ it follows that

$$\mathcal{N}(\lambda) = \mathcal{N}\big(m_{M^p_{g,\text{loc}}}\big) \quad \text{resp.} \quad \mathcal{N}_0(\lambda) = \mathcal{N}_0\big(m_{M^p_{g,\text{loc}}}\big).$$

Moreover, Theorem 3.3.1 guarantees that $L^1\big(m_{M^p_{g,\text{loc}}}\big)$ is the optimal domain of the multiplication operator $M^p_{g,\text{loc}}$ and the optimal extension of $M^p_{g,\text{loc}}$ is the integration operator $I_{m_{M^p_{g,\text{loc}}}} : L^1\big(m_{M^p_{g,\text{loc}}}\big) \to L^p_{\text{loc}}(\mathbb{R})$, which is given by

$$I_{m_{M^p_{g,\text{loc}}}}(f) = \int_{\mathbb{R}} f \, dm_{M^p_{g,\text{loc}}} \overset{(4.12)}{=} fg, \quad \text{for } f \in L^1\big(m_{M^p_{g,\text{loc}}}\big).$$

Before we continue our investigations on $L^1\big(m_{M^p_{g,\text{loc}}}\big)$ let us take a closer look at the vector space $\mathcal{M}^p_{\text{loc}}$.

It is clear that $L^\infty_{\text{loc}}(\mathbb{R}) \subseteq \mathcal{M}^p_{\text{loc}}$. Indeed, for any fixed $g \in L^\infty_{\text{loc}}(\mathbb{R})$ there exists, for every $k \in \mathbb{N}$, a constant $M_k > 0$ satisfying $|g(w)| \leqslant M_k$, for λ-almost every

$w \in [-k, k]$. Thus, for $f \in L^p_{\text{loc}}(\mathbb{R})$, we obtain that

$$q_k(fg) = \left(\int_{-k}^{k} |fg|^p \, d\lambda \right)^{1/p} \leqslant \left(\int_{-k}^{k} M_k^p |f|^p \, d\lambda \right)^{1/p}$$

$$= M_k \left(\int_{-k}^{k} |f|^p \, d\lambda \right)^{1/p}$$

$$= M_k \, q_k(f) < \infty,$$

for all $k \in \mathbb{N}$, which shows that $fg \in L^p_{\text{loc}}(\mathbb{R})$. Hence, in the case that $g \in L^\infty_{\text{loc}}(\mathbb{R})$, we can take up some results established in [1]. To do this, recall the definition of a spectral measure as given on page 80. This time we consider the spectral measure $\hat{P} : \mathcal{B}(\mathbb{R}) \to L_s\big(L^p_{\text{loc}}(\mathbb{R})\big)$ where, for each $A \in \mathcal{B}(\mathbb{R})$, the operator $\hat{P}(A)$ is given by

$$\hat{P}(A) : f \mapsto f\chi_A, \quad \text{for } f \in L^p_{\text{loc}}(\mathbb{R}).$$

In other words, $\hat{P}(A) = M^p_{\chi_A, \text{loc}}$ is the multiplication operator by the characteristic function χ_A. In accordance with the definition of integrability with respect to a spectral measure, [1, p. 101], a measurable function $g \in L^0(\mathbb{R})$ is $\hat{P}$-integrable if there exists an operator

$$\int_{\mathbb{R}} g \, d\hat{P} \in L\big(L^p_{\text{loc}}(\mathbb{R})\big)$$

such that g is integrable with respect to each complex measure

$$\langle \hat{P}f, \varphi \rangle : A \mapsto \langle \hat{P}(A)f, \varphi \rangle, \quad \text{for } A \in \mathcal{B}(\mathbb{R}),$$

where $f \in L^p_{\text{loc}}(\mathbb{R})$ and $\varphi \in \big(L^p_{\text{loc}}(\mathbb{R})\big)^*$, and such that

$$\left\langle \left(\int_{\mathbb{R}} g \, d\hat{P} \right) f, \varphi \right\rangle = \int_{\mathbb{R}} g \, d\langle \hat{P}f, \varphi \rangle,$$

for $f \in L^p_{\text{loc}}(\mathbb{R})$ and $\varphi \in \big(L^p_{\text{loc}}(\mathbb{R})\big)^*$. It turns out that each $g \in L^\infty_{\text{loc}}(\mathbb{R})$ is $\hat{P}$-integrable, even more: that $L^\infty_{\text{loc}}(\mathbb{R}) = L^1(\hat{P})$, and that the operator

$$I_{\hat{P}}(g) := \int_{\mathbb{R}} g \, d\hat{P} : f \mapsto fg, \quad \text{for } f \in L^p_{\text{loc}}(\mathbb{R}),$$

corresponds to the multiplication operator $M^p_{g, \text{loc}}$, [1, Proposition 17].

The question is whether the reverse inclusion $\mathcal{M}^p_{\text{loc}} \subseteq L^\infty_{\text{loc}}(\mathbb{R})$ also holds. It was already noted prior to Proposition 4.1.6 that $\mathcal{M}^p_{\text{loc}} \subseteq L^p_{\text{loc}}(\mathbb{R})$. But, since $L^\infty_{\text{loc}}(\mathbb{R}) \subseteq L^p_{\text{loc}}(\mathbb{R})$, for all $p \in (1, \infty)$, this does not mean that $\mathcal{M}^p_{\text{loc}}$ is necessarily strictly larger than $L^\infty_{\text{loc}}(\mathbb{R})$. To investigate this point further let $g \in \mathcal{M}^p_{\text{loc}}$. Then $fg \in L^p_{\text{loc}}(\mathbb{R})$, for all $f \in L^p_{\text{loc}}(\mathbb{R})$, meaning that

$$q_k(fg) = \left(\int_{-k}^{k} |fg|^p \, d\lambda \right)^{1/p} < \infty, \quad \text{for all } k \in \mathbb{N}.$$

Fix an arbitrary $k \in \mathbb{N}$. Denote by $\tilde{f}_k = f|_{[-k,k]}$ the restriction of any function $f \in L^p_{\text{loc}}(\mathbb{R})$ to the interval $[-k, k]$, by $\tilde{g}_k = g|_{[-k,k]}$ the restriction of g to $[-k, k]$ and by λ_k the restriction of λ to $[-k, k]$. Note that $L^p([-k, k])$, equipped with the norm

$$\|h\|_{p,k} := \left(\int_{-k}^{k} |h|^p \, d\lambda_k \right)^{1/p}, \quad \text{for } h \in L^p([-k, k]), \tag{4.13}$$

is a Banach space and that for $g \in \mathcal{M}^p_{\text{loc}}$ and $f \in L^p_{\text{loc}}(\mathbb{R})$ the equation

$$q_k(fg) = \left(\int_{-k}^{k} |fg|^p \, d\lambda \right)^{1/p} = \left(\int_{-k}^{k} |\tilde{f}_k \tilde{g}_k|^p \, d\lambda_k \right)^{1/p} = \|\tilde{f}_k \tilde{g}_k\|_{p,k}, \quad \text{for } k \in \mathbb{N},$$

holds. Since every $h \in L^p([-k, k])$ is of the form $h = \tilde{f}_k$ for some $f \in L^p_{\text{loc}}(\mathbb{R})$ it follows that

$$\|h\tilde{g}_k\|_{p,k} < \infty, \quad \text{for all } h \in L^p([-k, k]).$$

But, it is known that

$$\mathcal{M}\big(L^p([-k,k]), L^p([-k,k])\big) := \{h \in L^0([-k,k]) : h \, L^p([-k,k]) \subseteq L^p([-k,k])\}$$
$$= L^\infty([-k,k]),$$

[26, p. 47], and so $\tilde{g}_k \in L^\infty([-k, k])$. Thus, g has to be λ-essentially bounded on $[-k, k]$. As k was chosen arbitrarily we can conclude that $g \in L^\infty([-k, k])$, for all $k \in \mathbb{N}$, and consequently $g \in L^\infty_{\text{loc}}(\mathbb{R})$.

Putting together the previous discussion we obtain the following proposition.

Proposition 4.1.8
$\mathcal{M}^p_{loc} = L^\infty_{loc}(\mathbb{R}) = L^1(\hat{P})$, for every $p \in (1, \infty)$. $\quad\square$

We return to our investigation of the space $L^1\big(m_{M^p_{g,\text{loc}}}\big)$. As noted before, $L^p_{\text{loc}}(\mathbb{R}) \subseteq L^1\big(m_{M^p_{g,\text{loc}}}\big)$. The question is whether, for $g \in \mathcal{M}^p_{\text{loc}} = L^\infty_{\text{loc}}(\mathbb{R})$, the space $L^1\big(m_{M^p_{g,\text{loc}}}\big) = \{f \in L^0(\mathbb{R}) : fg \in L^p_{\text{loc}}(\mathbb{R})\}$ is strictly larger than $L^p_{\text{loc}}(\mathbb{R})$ or, in other words, whether there is a function $f \in L^0(\mathbb{R})$ satisfying $f \notin L^p_{\text{loc}}(\mathbb{R})$ but $fg \in L^p_{\text{loc}}(\mathbb{R})$.

It is not too difficult to find such a function when g is given by $g(w) := w$, for all $w \in \mathbb{R}$. Then g is certainly an element of $L^\infty_{\text{loc}}(\mathbb{R})$, since $|g| \leqslant k$ on $[-k, k]$, for every

91

$k \in \mathbb{N}$. Note that $g \neq 0$ λ-a.e. on $\mathbb{R}$ as well. Choose $f(w) := w^{-(1/p)}$, for all $w \in \mathbb{R}$. Then $f \notin L^p_{\mathrm{loc}}(\mathbb{R})$, since for any fixed $k \in \mathbb{N}$ we have

$$q^p_k(f) = \int_{-k}^{k} \left| w^{-(1/p)} \right|^p \, d\lambda = 2 \int_0^k w^{-1} \, d\lambda = 2 \lim_{\varepsilon \to 0} \left(\ln(k) - \ln(\varepsilon) \right) = \infty. \qquad (4.14)$$

But, when considering fg we obtain, for $k \in \mathbb{N}$ fixed, the equation

$$
\begin{aligned}
q^p_k(fg) &= \int_{-k}^{k} \left| w^{-(1/p)} \cdot w \right|^p \, d\lambda \\
&= \int_{-k}^{k} \left| w^{p-1} \right| \, d\lambda \\
&= \tfrac{1}{p} \left[w^p \right]_{-k}^{k} = \tfrac{1}{p} \left(k^p - (-1)^p k^p \right) < \infty
\end{aligned}
$$

yielding that $fg \in L^p_{\mathrm{loc}}(\mathbb{R})$.

Of course we can draw a line parallel to the results we obtained in Subsection 4.1.1 when investigating the multiplication operator on $L^{p-}([0,1])$. Since $M^p_{g,\mathrm{loc}}$ is λ-determined, $g \neq 0$ λ-a.e. on $\mathbb{R}$ and thus, any function $f \in L^1\left(m_{M^p_{g,\mathrm{loc}}} \right) = \{ f \in L^0(\mathbb{R}) : fg \in L^p_{\mathrm{loc}}(\mathbb{R}) \}$ can be written as

$$f = \tfrac{1}{g} \cdot \underbrace{gf}_{\in L^p_{\mathrm{loc}}(\mathbb{R})}.$$

If $\frac{1}{g} \in L^\infty_{\mathrm{loc}}(\mathbb{R}) = \mathcal{M}^p_{\mathrm{loc}}$ it is clear that $\frac{1}{g} \cdot gf \in L^p_{\mathrm{loc}}(\mathbb{R})$. The example above shows that this need not to be the case if $\frac{1}{g} \notin L^\infty_{\mathrm{loc}}(\mathbb{R})$. Indeed, for the function g we have chosen there $\frac{1}{g(w)} = \frac{1}{w}$ is not bounded on any of the intervals $[-k,k]$, for $k \in \mathbb{N}$.

Finally, we wish to know whether $m_{M^p_{g,\mathrm{loc}}}$ is of finite variation or not. To investigate this question let $p \in (1,\infty)$ and $g \in \mathcal{M}^p_{\mathrm{loc}}$. For each $k \in \mathbb{N}$ we define the local-Banach-space-valued vector measure $\left(m_{M^p_{g,\mathrm{loc}}} \right)_k : \mathcal{B}(\mathbb{R})|_{[-k,k]} \to L^p([-k,k])$ by

$$\left(m_{M^p_{g,\mathrm{loc}}} \right)_k := \chi_A \tilde{g}_k, \quad \text{for } A \in \mathcal{B}(\mathbb{R})|_{[-k,k]},$$

where $\tilde{g}_k$ is again the restriction of g to the interval $[-k,k]$. Fix $k \in \mathbb{N}$. Then we obtain the variation of $\left(m_{M^p_{g,\mathrm{loc}}} \right)_k$ by

$$
\begin{aligned}
\left| \left(m_{M^p_{g,\mathrm{loc}}} \right)_k \right|(A) &= \sup_\pi \sum_{j=1}^{l} \left\| \left(m_{M^p_{g,\mathrm{loc}}} \right)_k (A_j) \right\|_{p,k} \\
&= \sup_\pi \sum_{j=1}^{l} \left\| \chi_{A_j} \tilde{g}_k \right\|_{p,k}
\end{aligned}
$$

$$= \sup_{\pi} \sum_{j=1}^{l} \left(\int_{-k}^{k} |\chi_{A_j} \tilde{g}_k|^p \, d\lambda_k \right)^{1/p}$$

$$= \sup_{\pi} \sum_{j=1}^{l} \left(\int_{A_j \cap [-k,k]} |\tilde{g}_k|^p \, d\lambda_k \right)^{1/p} \tag{4.15}$$

where $A \in \mathcal{B}(\mathbb{R})|_{[-k,k]}$ and $\pi = \{A_j\}_{j=1}^{l}$ is any finite partition of A. Here, $\| \cdot \|_{p,k}$ is the norm in $L^p([-k,k])$ as defined in (4.13).

As in the case of M_g^{p-} it is easy to see that the variation of $\left(m_{M_{g,\mathrm{loc}}^p} \right)_k$ may not be finite. Indeed, choose $g := \chi_{\mathbb{R}}$. Then $g \in \mathcal{M}_{\mathrm{loc}}^p$, since

$$fg = f\chi_{\mathbb{R}} = f \in L_{\mathrm{loc}}^p(\mathbb{R}), \quad \text{for all } f \in L_{\mathrm{loc}}^p(\mathbb{R}).$$

Furthermore, $\tilde{g}_k := \chi_{[-k,k]}$, for all $k \in \mathbb{N}$. For $k, l \in \mathbb{N}$ fixed, consider the partition $\{A_j\}_{j=1}^{2l}$ of $[-k,k]$ defined by

$$A_j := \left[\frac{-k \cdot j}{l}, \frac{-k \cdot (j-1)}{l} \right), \quad \text{for } j = 1, \dots, l,$$

$$A_j := \left[\frac{k \cdot (j-l-1)}{l}, \frac{k \cdot (j-l)}{l} \right), \quad \text{for } j = l+1, \dots, 2l-1,$$

$$A_{2l} := \left[\frac{k \cdot (l-1)}{l}, k \right].$$

Since

$$\lambda_k(A_j \cap [-k,k]) = \frac{k}{l}, \quad \text{for all } j = 1, \dots, 2l,$$

we obtain that

$$\sum_{j=1}^{2l} \left(\int_{A_j \cap [-k,k]} |\tilde{g}_k|^p \, d\lambda_k \right)^{1/p} = \sum_{j=1}^{2l} \left(\lambda_k(A_j \cap [-k,k]) \right)^{1/p}$$

$$= \sum_{j=1}^{2l} \left(\tfrac{k}{l} \right)^{1/p} = 2 \, k^{1/p} \, l^{1-(1/p)} = 2 \, k^{1/p} \, l^{1/q}.$$

Here, q is the conjugate exponent of p and therefore satisfies $\frac{1}{q} > 0$. Letting $l \to \infty$ yields that $l^{1/q} \to \infty$ and we can conclude that, for $g = \chi_{\mathbb{R}}$, the variation of $\left(m_{M_{g,\mathrm{loc}}^p} \right)_k$ as given by (4.15) and thus, also of $m_{M_{g,\mathrm{loc}}^p}$ is infinite. Actually, this result does not depend on the function g.

Proposition 4.1.9

Let $g \in \mathcal{M}_{loc}^p \backslash \{0\}$. Then the variation of $m_{M_{g,loc}^p}$ is infinite.

Proof:

Let $g \in \mathcal{M}_{g,\mathrm{loc}}^p \backslash \{0\} = L_{\mathrm{loc}}^\infty(\mathbb{R}) \backslash \{0\}$. Then also $g \in L_{\mathrm{loc}}^p(\mathbb{R}) \backslash \{0\}$ and thus, there is an index $m \in \mathbb{N}$ such that $q_m(g) > 0$. Since $\{q_k\}_{k \in \mathbb{N}}$ is increasing it follows that $q_k(g) > 0$, for all $k \geqslant m$. Furthermore, for each $k \in \mathbb{N}$, $|g|^p|_{[-k,k]} \in L^1([-k,k])$. Hence, we can find an index $k \in \mathbb{N}$ and a set $A \in \mathcal{B}(\mathbb{R})|_{[-k,k]}$ such that

$$0 < \int_{A \cap [-k,k]} |g|^p \, d\lambda < \infty.$$

Fix such an $k \in \mathbb{N}$ and $A \in \mathcal{B}(\mathbb{R})|_{[-k,k]}$ and let

$$\int_{A \cap [-k,k]} |g|^p \, d\lambda =: \alpha.$$

Define, by using again the notations $\tilde{g}_k$ and λ_k for the restrictions of g and λ to $[-k,k]$ (see page 91), a set function $\nu : \mathcal{B}(A \cap [-k,k]) \to [0,\infty)$ by

$$\nu(B) := \int_B |g|^p \, d\lambda = \int_B |\tilde{g}_k|^p \, d\lambda_k, \quad \text{for } B \in \mathcal{B}(A \cap [-k,k]).$$

Since the Lebesgue measure is non-atomic on the real line, [9, p. 26], it follows that ν is also non-atomic on $\mathcal{B}(A \cap [-k,k])$. Fix $l \in \mathbb{N}$. Lemma 2.2.2 then implies that there exists a partition $\{A_j\}_{j=1}^l \subseteq \mathcal{B}(A \cap [-k,k])$ of A satisfying

$$\nu(A_j) = \int_{A_j} |\tilde{g}_k|^p \, d\lambda_k = \int_{A_j \cap [-k,k]} |\tilde{g}_k|^p \, d\lambda_k = \tfrac{\alpha}{l}, \quad \text{for all } j = 1, \ldots, l.$$

We finally obtain that

$$\sum_{j=1}^l \left(\int_{A_j \cap [-k,k]} |\tilde{g}_k|^p \, d\lambda_k \right)^{1/p} = \sum_{j=1}^l \left(\nu(A_j) \right)^{1/p}$$

$$= \sum_{j=1}^l \left(\tfrac{\alpha}{l} \right)^{1/p} = l \cdot \left(\tfrac{\alpha}{l} \right)^{1/p} = \alpha^{1/p} \, l^{1-(1/p)} = \alpha^{1/p} \, l^{1/q}$$

where q denotes the conjugate exponent of p and therefore satisfies $\frac{1}{q} > 0$. Letting $l \to \infty$ we get $\alpha^{1/p} \, l^{1/q} \to \infty$ and it follows that the variation of $\left(m_{M_{g,\mathrm{loc}}^p} \right)_k$ as given by (4.15) and thus, the variation of $m_{M_{g,\mathrm{loc}}^p}$ is infinite. $\quad \square$

4.2 The Volterra operator

Throughout this section we consider the finite measure space $([0,1], \mathcal{B}([0,1]), \lambda)$, where λ is Lebesgue measure and $\mathcal{B}([0,1])$ the σ-algebra of Lebesgue measurable subsets of $[0, 1]$. Let, for $p \in (1, \infty)$ fixed, $L^{p-} := L^{p-}([0, 1]) = \bigcap_{k \in \mathbb{N}} L^{r_k}([0, 1])$ with $1 \leqslant r_k \uparrow_k p$ be the Fréchet function space as discussed in Example 2.3.1. Furthermore, denote by $L^0 := L^0([0, 1])$ the Lebesgue measurable functions $f : [0, 1] \to \mathbb{C}$.

Define on L^{p-} the *Volterra operator* $V_{p-} : L^{p-} \to L^{p-}$ mapping $f \mapsto V_{p-}(f)$ where

$$V_{p-}(f)(w) := \int_0^w f(t) \, d\lambda(t), \quad \text{for } w \in [0, 1].$$

Remark 4.2.1

For each $f \in L^{p-}$, the function $V_{p-}(f) : [0, 1] \to \mathbb{C}$ is continuous on $[0, 1]$.

Proof of Remark 4.2.1:

Choose an arbitrary $f \in L^{p-}$. Fix $w_0 \in [0, 1]$ and let $\{w_n\}_{n \in \mathbb{N}} \subseteq [0, 1]$ be any sequence converging to w_0. Then it is clear that

$$\lim_{n \to \infty} f\chi_{[0, w_n]} = f\chi_{[0, w_0]}$$

pointwise on $[0, 1]$. Since $f \in L^{p-} \subseteq L^1$, also $|f| \in L^1$. Furthermore, $|f\chi_{[0, w_n]}| \leqslant |f|$, for all $n \in \mathbb{N}$. By applying Lebegue's Dominated Convergence Theorem 2.2.2 we obtain that

$$
\begin{aligned}
\lim_{n \to \infty} V_{p-}(f)(w_n) &= \lim_{n \to \infty} \int_0^{w_n} f(t) \, d\lambda(t) \\
&= \lim_{n \to \infty} \int_0^1 \left(f\chi_{[0, w_n]} \right)(t) \, d\lambda(t) \\
&= \int_0^1 \left(f\chi_{[0, w_0]} \right)(t) \, d\lambda(t) \\
&= \int_0^{w_0} f(t) \, d\lambda(t) = V_{p-}(f)(w_0)
\end{aligned}
$$

which shows that $V_{p-}(f)$ is a continuous function. $\square$

The linearity of the Volterra operator results from the linearity of the Lebesgue integral. Let us state here two further properties of the Volterra operator.

Proposition 4.2.1

The Volterra operator $V_{p-} : L^{p-} \to L^{p-}$ is continuous.

Proof:

Let $f \in L^{p-}$ be arbitrarily chosen. Then $f \in L^{r_1} \subseteq L^1$ and so, for each $w \in [0, 1]$,

95

we have, by Hölder's inequality, that

$$
\begin{aligned}
|V_{p-}(f)(w)| &= \left| \int_0^w f(t)\, d\lambda(t) \right| \\
&\leqslant \int_0^w |f(t)|\, d\lambda(t) \\
&= \int_0^1 |f(t)\chi_{[0,w]}(t)|\, d\lambda(t) \\
&\overset{(2.9)}{\leqslant} \left(\int_0^1 |f(t)|^{r_1}\, d\lambda(t) \right)^{1/r_1} \left(\int_0^1 |\chi_{[0,w]}(t)|^{s_1}\, d\lambda(t) \right)^{1/s_1} \\
&= q_1(f)\, \lambda([0,w])^{1/s_1} \leqslant q_1(f),
\end{aligned}
$$

where s_1 is the conjugate exponent of r_1. Hence, for each $k \in \mathbb{N}$, we obtain that

$$
\begin{aligned}
q_k\big(V_{p-}(f)\big) &= \left(\int_0^1 |V_{p-}(f)|^{r_k}\, d\lambda \right)^{1/r_k} \\
&\leqslant \left(\int_0^1 (q_1(f))^{r_k}\, d\lambda \right)^{1/r_k} \\
&= q_1(f)\, \lambda([0,1])^{1/r_k},
\end{aligned}
$$

which implies that V_{p-} is continuous. $\quad\square$

Proposition 4.2.2

The Volterra operator $V_{p-} : L^{p-} \to L^{p-}$ is injective.

Proof:

Let $f \in L^{p-}$ satisfy $V_{p-}(f) = 0$ meaning that

$$
\int_0^w f(t)\, d\lambda(t) = 0, \quad \text{for all } w \in [0,1].
$$

The Fundamental Theorem of Calculus, [11, p. 304], then yields that

$$
f(w) = \left(\int_0^w f(t)\, d\lambda(t) \right)' = 0' = 0, \quad \text{for } \lambda\text{-almost every } w \in [0,1].
$$

Thus, V_{p-} is injective. $\quad\square$

In the following pages we are going to study the vector measure $m_{V_{p-}}$ associated with the operator V_{p-}; more explicitly, the vector measure $m_{V_{p-}} : \mathcal{B}([0,1]) \to L^{p-}$ defined by

$$
m_{V_{p-}}(A) := V_{p-}(\chi_A)
$$

where, for each $A \in \mathcal{B}([0, 1])$,

$$V_{p-}(\chi_A)(w) = \int_0^w \chi_A(t) \, d\lambda(t) = \lambda\big([0, w] \cap A\big), \quad \text{for } w \in [0, 1].$$

Since L^{p-} contains the $\mathcal{B}([0, 1])$-simple functions and has a σ-Lebesgue topology, $m_{V_{p-}}$ is by Proposition 3.2.1 indeed a vector measure. For each $\varphi \in \big(L^{p-}\big)^* = \bigcup_{k \in \mathbb{N}} L^{s_k}$ with $\frac{1}{r_k} + \frac{1}{s_k} = 1$, for $k \in \mathbb{N}$, we induce the scalar measure $\langle m_{V_{p-}}, \varphi \rangle :$ $\mathcal{B}([0, 1]) \to \mathbb{C}$ given by

$$\langle m_{V_{p-}}, \varphi \rangle(A) := \langle m_{V_{p-}}(A), \varphi \rangle, \quad \text{for } A \in \mathcal{B}([0, 1]).$$

By applying the identity $\chi_{[0,w]}(t) = \chi_{[t,1]}(w)$ and Fubini's Theorem 2.2.3, this scalar measure can also be expressed by the following term:

$$
\begin{aligned}
\langle m_{V_{p-}}(A), \varphi \rangle &= \int_0^1 V_{p-}(\chi_A)(w) \, \varphi(w) \, d\lambda(w) \\
&= \int_0^1 \left(\int_0^w \chi_A(t) \, d\lambda(t) \right) \varphi(w) \, d\lambda(w) \\
&= \int_0^1 \left(\int_0^1 \chi_{[0,w]}(t) \, \chi_A(t) \, d\lambda(t) \right) \varphi(w) \, d\lambda(w) \\
&= \int_0^1 \left(\int_0^1 \chi_{[t,1]}(w) \, \chi_A(t) \, d\lambda(t) \right) \varphi(w) \, d\lambda(w) \\
&= \int_0^1 \chi_A(t) \left(\int_0^1 \chi_{[t,1]}(w) \, \varphi(w) \, d\lambda(w) \right) d\lambda(t) \\
&= \int_A \langle \chi_{[t,1]}, \varphi \rangle \, d\lambda(t),
\end{aligned}
\tag{4.16}
$$

for $A \in \mathcal{B}([0, 1])$. Observe that $t \mapsto \langle \chi_{[t,1]}, \varphi \rangle \in L^1$ since

$$
\begin{aligned}
\int_0^1 \big| \langle \chi_{[t,1]}, \varphi \rangle \big| \, d\lambda(t) &= \int_0^1 \left| \int_0^1 \chi_{[t,1]}(w) \, \varphi(w) \, d\lambda(w) \right| d\lambda(t) \\
&\leqslant \int_0^1 \left(\int_t^1 |\varphi(w)| \, d\lambda(w) \right) d\lambda(t) \\
&\leqslant \int_0^1 \left(\int_0^1 |\varphi(w)| \, d\lambda(w) \right) d\lambda(t) \\
&= \|\varphi\|_1 \, \lambda([0, 1]) \; < \; \infty,
\end{aligned}
$$

because $\varphi \in L^{s_k}$ for some $k \in \mathbb{N}$ and so $\varphi \in L^1$. Hence, by Proposition 2.2.4, the variation of $\langle m_{V_{p-}}, \varphi \rangle$ is given by

$$\big| \langle m_{V_{p-}}, \varphi \rangle \big|(A) = \int_A \big| \langle \chi_{[t,1]}, \varphi \rangle \big| \, d\lambda(t), \quad \text{for } A \in \mathcal{B}([0, 1]).
\tag{4.17}$$

One of the main objects of interest will be the space of $m_{V_{p-}}$-integrable functions. Recall that a function $f : [0,1] \to \mathbb{C}$ is $m_{V_{p-}}$-integrable if it is scalarly $m_{V_{p-}}$-integrable, i.e., if it is integrable with respect to each scalar measure $\langle m_{V_{p-}}, \varphi \rangle$, for $\varphi \in \left(L^{p-} \right)^*$, and if, for each $A \in \mathcal{B}([0,1])$, there exists an element $\int_A f \, dm_{V_{p-}} \in L^{p-}$ satisfying

$$\left\langle \int_A f \, dm_{V_{p-}}, \varphi \right\rangle = \int_A f \, d\langle m_{V_{p-}}, \varphi \rangle,$$

for all $\varphi \in \left(L^{p-} \right)^*$. Define

$$h(t) := \chi_{[t,1]}, \quad \text{for } t \in [0,1], \tag{4.18}$$

and observe that $h(t) \in \mathrm{sim}\big(\mathcal{B}([0,1])\big) \subseteq L^{p-}$, for each $t \in [0,1]$. Then (4.16) becomes

$$\langle m_{V_{p-}}(A), \varphi \rangle = \int_A \langle h(t), \varphi \rangle \, d\lambda(t), \quad \text{for } A \in \mathcal{B}([0,1]).$$

Moreover,

$$h(t)(w) = \chi_{[t,1]}(w) = \chi_{[0,w]}(t),$$

for all $t, w \in [0,1]$. Taking a closer look at the previous equations we derive, for $\varphi \in \left(L^{p-} \right)^*$, $f \in L^1(m_{V_{p-}})$ and $A \in \mathcal{B}([0,1])$, after noting that also $|f| \in L^1(m_{V_{p-}})$ and $|\varphi| \in \left(L^{p-} \right)^*$, that

$$
\begin{aligned}
&\int_A |f| \, d\langle m_{V_{p-}}, |\varphi| \rangle \\
&\overset{(4.16)}{=} \int_A |f(t)| \, \langle \chi_{[t,1]}, |\varphi| \rangle \, d\lambda(t) \\
&= \int_0^1 \chi_A(t) \, |f(t)| \left(\int_0^1 \chi_{[t,1]}(w) \, |\varphi(w)| \, d\lambda(w) \right) d\lambda(t) \\
&= \int_0^1 \chi_A(t) \, |f(t)| \left(\int_0^1 \chi_{[0,w]}(t) \, |\varphi(w)| \, d\lambda(w) \right) d\lambda(t) \\
&= \int_0^1 \left(\int_0^1 \chi_A(t) \, |f(t)| \, \chi_{[0,w]}(t) \, d\lambda(t) \right) |\varphi(w)| \, d\lambda(w). \tag{4.19}
\end{aligned}
$$

Define the function

$$g_{A,f}(w) := \int_A f(t) \, h(t)(w) \, d\lambda(t) = \int_0^w \chi_A(t) \, f(t) \, d\lambda(t), \quad \text{for } w \in [0,1].$$

We need to prove that $g_{A,f} \in L^{p-}$. Since $|g_{A,f}| \leqslant g_{A,|f|}$, with $|f| \in L^1(m_{V_{p-}})$, it suffices to show that $g_{A,|f|} \in L^{p-}$. Fix $k \in \mathbb{N}$ and let $\varphi \in L^{s_k} \subseteq \left(L^{p-} \right)^*$ with

$\frac{1}{r_k} + \frac{1}{s_k} = 1$. Then also $|\varphi| \in L^{s_k}$ and the measure $\langle m_{V_{p-}}, |\varphi| \rangle \geqslant 0$ is given by

$$\langle m_{V_{p-}}, |\varphi| \rangle(A) = \int_A \langle \chi_{[t,1]}, |\varphi| \rangle \, d\lambda(t), \quad \text{for } A \in \mathcal{B}([0,1]);$$

see (4.16). By (4.19) we have

$$\int_0^1 g_{A,|f|}(w) \, |\varphi(w)| \, d\lambda(w) = \int_A |f| \, d\langle m_{V_{p-}}, |\varphi| \rangle < \infty. \qquad (4.20)$$

Since L^{s_k} is a reflexive Banach space and (4.20) holds for all $\varphi \in L^{s_k}$, it follows that $g_{A,|f|} \in L^{r_k}$. But, this is valid for all $k \in \mathbb{N}$ and so $g_{A,|f|} \in L^{p-}$, as was to be proved. Hence, the indefinite integral of $f \in L^1(m_{V_{p-}})$ over $A \in \mathcal{B}([0,1])$ can be expressed by

$$\int_A f \, dm_{V_{p-}} = \int_A f \chi_{[0,\cdot]} \, d\lambda = \int_A f(t) \, h(t)(\cdot) \, d\lambda \qquad (4.21)$$

and the set function $m_{V_{p-},f} : \mathcal{B}([0,1]) \to L^{p-}$ associated with the indefinite integral is given by

$$m_{V_{p-},f}(A) := \int_A f \, dm_{V_{p-}} = \int_A f(t) \, h(t)(\cdot) \, d\lambda.$$

By the Orlicz-Pettis Theorem 2.1.3 it is again a vector measure.

Before we begin examining the space $L^1(m_{V_{p-}})$ let us state here some conclusions we can draw immediately from the properties of L^{p-} and of the operator V_{p-}. Proposition 3.2.2 ensures that each $f \in L^{p-}$ is $m_{V_{p-}}$-integrable, i.e., that

$$L^{p-} \subseteq L^1(m_{V_{p-}}),$$

and that, for each $f \in L^{p-}$, the equation

$$V_{p-}(f\chi_A) = \int_A f(t) \, h(t)(\cdot) \, d\lambda = \int_A f \, dm_{V_{p-}}, \quad \text{for } A \in \mathcal{B}([0,1]),$$

holds. Furthermore, since V_{p-} is injective (by Proposition 4.2.2), we can conclude by Corollary 3.2.1 that V_{p-} is λ-determined which, on the other hand, implies that the λ-null functions and the $m_{V_{p-}}$-null functions coincide, i.e., that

$$\mathcal{N}(\lambda) = \mathcal{N}(m_{V_{p-}}),$$

and according to Lemma 3.2.2 the same is true for the null sets, i.e.,

$$\mathcal{N}_0(\lambda) = \mathcal{N}_0(m_{V_{p-}}).$$

Moreover, Theorem 3.3.1 yields that $L^1(m_{V_{p-}})$ is the optimal domain of the operator V_{p-} and the optimal extension of V_{p-} is the integration operator $I_{m_{V_{p-}}} : L^1(m_{V_{p-}}) \to L^{p-}$ given by

$$I_{m_{V_{p-}}}(f) := \int_0^1 f \, dm_{V_{p-}} \overset{(4.21)}{=} \int_0^1 f(t) \, h(t)(\cdot) \, d\lambda. \tag{4.22}$$

Another important property of the space L^{p-} is that it is reflexive which ensures that each scalarly $m_{V_{p-}}$-integrable function $f \in L^1_w(m_{V_{p-}})$ is already $m_{V_{p-}}$-integrable, meaning that $L^1(m_{V_{p-}}) = L^1_w(m_{V_{p-}})$.

In the forthcoming investigations we want to find answers to the following two questions: Is the inclusion $L^{p-} \subseteq L^1(m_{V_{p-}})$ proper? Is it possible to characterize the space $L^1(m_{V_{p-}})$? From the above discussion we see that $f \in L^1(m_{V_{p-}})$ if and only if the function $w \mapsto \int_A f(t) \chi_{[t,1]}(w) \, d\lambda(t)$, for $w \in [0,1]$, belongs to L^{p-} for every $A \in \mathcal{B}([0,1])$.

Although it is not a characterization of $L^1(m_{V_{p-}})$, the following result exhibits another space of functions which is contained in $L^1(m_{V_{p-}})$.

Proposition 4.2.3

Fix $p \in (1, \infty)$. Let $f : [0,1] \to [0,\infty)$ be a function which is Lebesgue integrable over $[0,w]$, for each $w \in [0,1)$, and such that the function $F_f : [0,1) \to \mathbb{R}$ defined by

$$F_f(w) := \int_0^w f(t) \, d\lambda(t)$$

is an element of L^{p-}. Then f is $m_{V_{p-}}$-integrable.

Proof:

Let f be as in the statement of the proposition. To show the $m_{V_{p-}}$-integrability of f let $\{s_n\}_{n \in \mathbb{N}} \subseteq \text{sim}(\mathcal{B}([0,1]))$ be any sequence of $\mathcal{B}([0,1])$-simple functions satisfying $0 \leqslant s_n \uparrow_n f$ pointwise on $[0,1]$. Fix $A \in \mathcal{B}([0,1])$. Then also $0 \leqslant s_n \chi_A \uparrow_n f \chi_A$ pointwise on $[0,1]$. Furthermore, since f is Lebesgue integrable over $[0,w]$, for each $w \in [0,1)$, the same is true for $f \chi_A$ and, as $F_f \in L^{p-}$, with L^{p-} a Fréchet function space, the function $F_{f,A} : [0,1) \to \mathbb{R}$ defined by

$$F_{f,A}(w) := \int_0^w f(t) \chi_A(t) \, d\lambda(t)$$

is an element of $L^{p-} \subseteq L^1(m_{V_{p-}})$ as well as it satisfies $F_{f,A} \leqslant F_f$ pointwise λ-a.e.. Moreover, from (4.16) and the definition of the integral it follows that $\int_A s_n \, dm_{V_{p-}}$

is the function

$$\int_A s_n \, dm_{V_{p-}} : w \mapsto \int_0^w s_n(t) \, \chi_A(t) \, d\lambda(t), \quad \text{for } w \in [0,1).$$

The calculation (4.19) reveals that $\int_A F_f \, dm_{V_{p-}}$ is the function

$$\int_A F_f \, dm_{V_{p-}} : w \mapsto \int_0^w f(t) \, \chi_A(t) \, d\lambda(t) = F_{f,A}(w), \quad \text{for } w \in [0,1).$$

By applying the Monotone Convergence Theorem 2.2.1 on $[0,w]$ for each fixed $0 \leqslant w < 1$ we obtain that

$$\lim_{n \to \infty} \int_0^w s_n(t) \, \chi_A(t) \, d\lambda(t) = F_{f,A}(w). \tag{4.23}$$

But, as L^{p-} has a σ-Lebesgue topology and since (4.23) is equivalent to $\left(F_{f,A} - \int_A s_n \, dm_{V_{p-}} \right) \downarrow_n 0$ pointwise, the sequence $\left\{ \int_A s_n \, dm_{V_{p-}} \right\}_{n \in \mathbb{N}} \subseteq L^{p-}$ converges to $F_{f,A}$ in the topology of L^{p-}. It follows from Proposition 2.4.1 that f is then $m_{V_{p-}}$-integrable. $\square$

A full characterization of the space $L^1(m_{V_{p-}})$ will be given later in Lemma 4.2.3. However, the first question concerning the inclusion $L^{p-} \subseteq L^1(m_{V_{p-}})$ can be answered without a major effort.

Proposition 4.2.4
Let $f \in L^1$. Then f is $m_{V_{p-}}$-integrable.

Proof:

Let $f \in L^1$. Then,

$$\|f\|_1 = \int_0^1 |f(t)| \, d\lambda(t) < \infty.$$

Fix $k \in \mathbb{N}$ and choose an arbitrary $A \in \mathcal{B}([0,1])$. Then,

$$
\begin{aligned}
q_k \left(\int_A f(t) \, h(t)(\cdot) \, d\lambda \right) &= \left(\int_0^1 \left| \int_0^w f(t) \, \chi_A(t) \, d\lambda(t) \right|^{r_k} d\lambda(w) \right)^{1/r_k} \\
&\leqslant \left(\int_0^1 \left(\int_0^w |f(t)| \, \chi_A(t) \, d\lambda(t) \right)^{r_k} d\lambda(w) \right)^{1/r_k} \\
&\leqslant \left(\int_0^1 \left(\int_0^1 |f(t)| \, d\lambda(t) \right)^{r_k} d\lambda(w) \right)^{1/r_k} \\
&= \left(\int_0^1 \|f\|_1^{r_k} \, d\lambda \right)^{1/r_k}
\end{aligned}
$$

101

$$= \|f\|_1 \underbrace{\lambda([0,1])^{1/r_k}}_{=1} = \|f\|_1 < \infty,$$

meaning that $\int_A f(t)\,h(t)(\cdot)\,d\lambda \in L^{p-}$. As $A \in \mathcal{B}([0,1])$ was chosen arbitrarily,

$$q_k(m_{V_{p-}})(f) = \sup\left\{ q_k\left(\int_A f(t)\,h(t)(\cdot)\,d\lambda\right) : A \in \mathcal{B}([0,1]) \right\} \leqslant \|f\|_1 < \infty.$$

This is true for all $k \in \mathbb{N}$ implying that f is $m_{V_{p-}}$-integrable, i.e., $f \in L^1(m_{V_{p-}})$.
$\square$

From (2.7) and Proposition 4.2.4 we can deduce the following chain of inclusions

$$L^{p-} \subsetneq L^1 \subseteq L^1(m_{V_{p-}})$$

where the first inclusion is indeed proper. To see this, consider the function $f : (0,1] \to \mathbb{C}$ defined by

$$f(w) := w^{-1/r}, \quad \text{for } w \in (0,1],$$

where $r > 1$ is an element of the sequence $\{r_k\}_{k\in\mathbb{N}} \subseteq \mathbb{R}$ satisfying $1 \leqslant r_k \uparrow_k p$ and $L^{p-} = \bigcap_{k\in\mathbb{N}} L^{r_k}$. Then $f \in L^1$, since

$$\int_0^1 \left| w^{-1/r} \right| d\lambda(w) = \left[\frac{1}{1 - \frac{1}{r}}\, w^{1-(1/r)} \right]_0^1 = \left[s\, w^{1/s} \right]_0^1 = s,$$

where s is the conjugate exponent of r. However, $f \notin L^{p-}$ as $f \notin L^r$ because

$$\int_0^1 \left| w^{-1/r} \right|^r d\lambda(w) = \int_0^1 w^{-1}\, d\lambda(w) = \lim_{\varepsilon \to 0} \left(\ln(1) - \ln(\varepsilon) \right) = \infty.$$

Hence, $L^{p-} \subsetneq L^1$. The question arises whether the inclusion $L^1 \subseteq L^1(m_{V_{p-}})$ is proper as well. To answer this question let us prove some further lemmas.

Define the L^{p-}-valued function $g_{p-} : [0,1] \to L^{p-}$ by

$$g_{p-}(t) := \chi_{[t,1]}. \tag{4.24}$$

Of course, this is the function h in (4.18) but now we need to examine the dependence of this function on p. Due to its "nice" properties the function g_{p-} will play a major role in the forthcoming investigations.

Lemma 4.2.1

For each $p \in (1, \infty)$, the function g_{p-} as defined in (4.24) is Bochner λ-integrable.

Proof:

Let $s, t \in [0, 1]$ be any pair of distinct real numbers. Without loss of generality, assume that $s < t$. For each $k \in \mathbb{N}$ we thereby obtain that

$$
\begin{aligned}
q_k\big(g_{p-}(s) - g_{p-}(t)\big) &= q_k\big(\chi_{[s,1]} - \chi_{[t,1]}\big) \\
&= q_k\big(\chi_{[s,t]}\big) \\
&= \left(\int_0^1 |\chi_{[s,t]}|^{r_k} \, d\lambda\right)^{1/r_k} \\
&= \lambda\big([s,t]\big)^{1/r_k} = |t - s|^{1/r_k},
\end{aligned}
$$

which shows that g_{p-} is a continuous function.

As $[0, 1]$ is compact and g_{p-} is continuous, it follows that $g_{p-}([0, 1])$ is compact in the Fréchet space L^{p-} meaning that g_{p-} is a compact metric space and thus, is separable, [22, pp. 18–19]. Since $g_{p-}([0, 1])$ is a Suslin space we can make use of [34, pp. 67–68] to conclude that g_{p-} is strongly λ-measurable.

Furthermore, we have for each $k \in \mathbb{N}$

$$
q_k\big(g_{p-}(t)\big) = q_k\big(\chi_{[t,1]}\big) = \left(\int_0^1 |\chi_{[t,1]}|^{r_k} d\lambda\right)^{1/r_k} = \lambda([t,1])^{1/r_k} = (1 - t)^{1/r_k} \quad (4.25)
$$

and therefore obtain that

$$
\int_0^1 q_k\big(g_{p-}(t)\big) \, d\lambda(t) \stackrel{(4.25)}{=} \int_0^1 (1 - t)^{1/r_k} \, d\lambda(t) = \frac{r_k}{1 + r_k} < \infty. \quad (4.26)
$$

Thus, g_{p-} is Bochner λ-integrable. $\quad\square$

It turns out that the Bochner λ-integral of g_{p-} coincides with $m_{V_{p-}}$, since for any fixed $A \in \mathcal{B}([0, 1])$ we have

$$
(B) - \int_A g_{p-}(t) \, d\lambda(t) = (B) - \int_A \chi_{[t,1]} \, d\lambda(t)
$$

and so, for $\varphi \in \left(L^{p-}\right)^*$,

$$
\begin{aligned}
\left\langle (B) - \int_A g_{p-}(t) \, d\lambda(t), \varphi \right\rangle &\stackrel{(2.36)}{=} \int_A \langle \chi_{[t,1]}, \varphi \rangle \, d\lambda(t) \\
&= \int_A \left(\int_0^1 \chi_{[t,1]}(w) \, \varphi(w) \, d\lambda(w)\right) d\lambda(t)
\end{aligned}
$$

$$\overset{(4.16)}{=} \ \langle m_{V_{p-}}(A), \varphi \rangle .$$

Since $\varphi \in \left(L^{p-} \right)^*$ is arbitrary, we can conclude that

$$(B) - \int_A g_{p-}(t)\, d\lambda(t) = m_{V_{p-}}(A),$$

for all $A \in \mathcal{B}([0,1])$. This result gives a new insight into the theory of the vector measure associated with the Volterra operator; it also gives the way to further investigations. We recall from Section 2.4 that the indefinite Bochner λ-integral $\lambda_{g_{p-}} : \mathcal{B}([0,1]) \to L^{p-}$ defined by

$$\lambda_{g_{p-}}(A) := (B) - \int_A g_{p-}(t)\, d\lambda(t), \quad \text{for } A \in \mathcal{B}([0,1]),$$

is an L^{p-}-valued vector measure of finite variation. Moreover, for each $k \in \mathbb{N}$, the variation $|(\lambda_{g_{p-}})_k|$ is given by

$$|(\lambda_{g_{p-}})_k|(A) = \int_A q_k\big(g_{p-}(t)\big)\, d\lambda(t) \overset{(4.25)}{=} \int_A (1-t)^{1/r_k}\, d\lambda(t),$$

for $A \in \mathcal{B}([0,1])$. We therefore have, for each $k \in \mathbb{N}$, that the local-Banach-space-valued vector measure $(m_{V_{p-}})_k : \mathcal{B}([0,1]) \to L^{r_k}$ is given by

$$(m_{V_{p-}})_k(A) = \int_A g_{r_k}(t)\, d\lambda(t), \quad \text{for } A \in \mathcal{B}([0,1]), \tag{4.27}$$

where again $g_{r_k}(t) := \chi_{[t,1]}$, for all $t \in [0,1]$, but this time considered as being an L^{r_k}-valued function. Its variation measure is given by

$$|(m_{V_{p-}})_k|(A) = |(\lambda_{g_{p-}})_k|(A) = \int_A (1-t)^{1/r_k}\, d\lambda(t), \quad \text{for } A \in \mathcal{B}([0,1]);$$

see [28, Section 5]. Further investigations concerning the vector measure $(m_{V_{p-}})_k$ and its variation also occur in [28]. As a consequence there is another interesting space of integrable functions to investigate, namely

$$\bigcap_{k \in \mathbb{N}} L^1\big(|(m_{V_{p-}})_k|\big) = \bigcap_{k \in \mathbb{N}} L^1\big((1-t)^{1/r_k}\, d\lambda(t)\big).$$

Recall that the topology of this Fréchet space is generated by the increasing sequence of semi-norms $\{\, \rho_k \,\}_{k \in \mathbb{N}}$ where, for each $k \in \mathbb{N}$,

$$\rho_k(f) := \int_0^1 |f|\, d|(m_{V_{p-}})_k| = \int_0^1 |f(t)|\, (1-t)^{1/r_k}\, d\lambda(t) < \infty, \tag{4.28}$$

for $f \in \bigcap_{k \in \mathbb{N}} L^1\big(|(m_{V_{p-}})_k|\big)$; see Section 2.4.

In the forthcoming investigations we will make use of the following two lemmata. For the Banach space setting of L^p they occur as Lemma 5.1 and Lemma 5.3 in [28].

Lemma 4.2.2

Let $1 < p < \infty$ with g_{p-} as defined in (4.24) and let $f : [0,1] \to \mathbb{C}$ be a measurable function. Then the following assertions are equivalent:

(i) $f g_{p-}$ is Bochner λ-integrable as an L^{p-}-valued function.

(ii) $\int_0^1 |f(t)|\,(1-t)^{1/r_k}\,d\lambda(t) < \infty$, for all $k \in \mathbb{N}$.

(iii) $f \in \bigcap_{k \in \mathbb{N}} L^1\big(|(m_{V_{p-}})_k|\big)$.

(iv) The function F_f given by

$$F_f(w) := \int_0^w |f(t)|\,d\lambda(t)$$

is defined for λ-almost every $w \in [0,1]$ and, for each $k \in \mathbb{N}$, the function $G_{f,k}$ given by

$$G_{f,k}(w) := (1-w)^{-1/s_k} F_f(w)$$

(where s_k is the conjugate exponent of r_k) is an element of L^1.

Proof:

(i) $\Leftrightarrow$ (ii) is clear as, for each $k \in \mathbb{N}$, the following equality holds:

$$
\begin{aligned}
\int_0^1 q_k\big((f g_{p-})(t)\big)\,d\lambda(t)
&= \int_0^1 \left(\int_0^1 \big| f(t)\,g_{p-}(t)(w) \big|^{r_k}\,d\lambda(w) \right)^{1/r_k} d\lambda(t) \\
&= \int_0^1 |f(t)| \left(\int_0^1 \big| g_{p-}(t)(w) \big|^{r_k}\,d\lambda(w) \right)^{1/r_k} d\lambda(t) \\
&= \int_0^1 |f(t)|\, q_k\big(g_{p-}(t)\big)\,d\lambda(t) \\
&\overset{(4.25)}{=} \int_0^1 |f(t)|\,(1-t)^{1/r_k}\,d\lambda(t).
\end{aligned}
$$

So, the left-side is finite for each $k \in \mathbb{N}$ (that is, $f g_{p-}$ is Bochner λ-integrable) if and only if the right-side is finite for each $k \in \mathbb{N}$.

(ii) $\Leftrightarrow$ (iii) results from the definition of the semi-norms in $\bigcap_{k \in \mathbb{N}} L^1\big(|(m_{V_{p-}})_k|\big)$. For each $k \in \mathbb{N}$, we have

$$\rho_k(f) = \int_0^1 |f|\,d|(m_{V_{p-}})_k| \overset{(4.28)}{=} \int_0^1 |f(t)|\,(1-t)^{1/r_k}\,d\lambda(t).$$

(iii) $\Leftrightarrow$ (iv) To prove this equivalence fix an arbitrary $k \in \mathbb{N}$ and let s_k be the conjugate exponent of r_k. Then $\frac{1}{r_k} = 1 - \frac{1}{s_k}$ and, for each $t \in [0, 1]$, the equality

$$\int_t^1 (1-w)^{-1/s_k}\, d\lambda(w) = \left[-\left(1-\tfrac{1}{s_k}\right)^{-1}(1-w)^{1-1/s_k}\right]_t^1 = r_k\,(1-t)^{1/r_k} \quad (4.29)$$

holds. By applying Fubini's Theorem 2.2.3 we obtain that

$$
\begin{aligned}
r_k \int_0^1 |f|\, d|(m_{V_{p-}})_k| \;&\overset{(4.28)}{=}\; r_k \int_0^1 |f(t)|\,(1-t)^{1/r_k}\, d\lambda(t) \\
&\overset{(4.29)}{=}\; \int_0^1 |f(t)| \left(\int_t^1 (1-w)^{-1/s_k}\, d\lambda(w)\right) d\lambda(t) \\
&=\; \int_0^1 |f(t)| \left(\int_0^1 \chi_{[t,1]}(w)\,(1-w)^{-1/s_k}\, d\lambda(w)\right) d\lambda(t) \\
&=\; \int_0^1 |f(t)| \left(\int_0^1 \chi_{[0,w]}(t)\,(1-w)^{-1/s_k}\, d\lambda(w)\right) d\lambda(t) \\
&=\; \int_0^1 (1-w)^{-1/s_k} \left(\int_0^1 \chi_{[0,w]}(t)\,|f(t)|\, d\lambda(t)\right) d\lambda(w) \\
&=\; \int_0^1 (1-w)^{-1/s_k} \left(\int_0^w |f(t)|\, d\lambda(t)\right) d\lambda(w) \\
&=\; \int_0^1 (1-w)^{-1/s_k} F_f(w)\, d\lambda(w) \\
&=\; \int_0^1 G_{f,k}(w)\, d\lambda(w).
\end{aligned}
$$

Hence, $\int_0^1 |f|\, d|(m_{V_{p-}})_k| < \infty$, for all $k \in \mathbb{N}$ (that is, $f \in \bigcap_{k \in \mathbb{N}} L^1(|(m_{V_{p-}})_k|)$), if and only if $G_{f,k} \in L^1$, for all $k \in \mathbb{N}$. $\quad\square$

Lemma 4.2.3

Let $1 < p < \infty$ with g_{p-} as defined in (4.24) and let $f : [0,1] \to \mathbb{C}$ be a measurable function. Then the following assertions are equivalent:

(i) $f g_{p-}$ is Pettis λ-integrable as an L^{p-}-valued function.

(ii) $f \in L^1(m_{V_{p-}})$.

(iii) $\int_0^1 |f(t)|\, |\langle g_{p-}(t), \varphi\rangle|\, d\lambda(t) < \infty$, for every $\varphi \in (L^{p-})^*$.

(iv) The function F_f as given in Lemma 4.2.2 is defined for λ-almost every $w \in [0,1]$ and $F_f \in L^{p-}$.

Proof:

(i) $\Leftrightarrow$ (iii) follows from the reflexivity of L^{p-} and the equalities

$$\int_0^1 \left|\langle (fg_{p-})(t), \varphi \rangle\right| d\lambda(t) = \int_0^1 \left|f(t) \langle g_{p-}(t), \varphi \rangle\right| d\lambda(t)$$
$$= \int_0^1 |f(t)| \left|\langle g_{p-}(t), \varphi \rangle\right| d\lambda(t),$$

for each $\varphi \in (L^{p-})^*$.

(ii) $\Leftrightarrow$ (iii) Using formula (4.16) the scalar measure $\langle m_{V_{p-}}, \varphi \rangle$ can, for each $\varphi \in (L^{p-})^*$, be written as

$$\langle m_{V_{p-}}, \varphi \rangle(A) \stackrel{(4.16)}{=} \int_A \langle \chi_{[t,1]}, \varphi \rangle d\lambda(t) = \int_A \langle g_{p-}(t), \varphi \rangle d\lambda(t),$$

for all $A \in \mathcal{B}([0,1])$. Furthermore, by (4.17), the variation of $\langle m_{V_{p-}}, \varphi \rangle$ is given by

$$\left|\langle m_{V_{p-}}, \varphi \rangle\right|(A) = \int_A \left|\langle g_{p-}(t), \varphi \rangle\right| d\lambda(t), \quad \text{for } A \in \mathcal{B}([0,1]).$$

Since $L^1(m_{V_{p-}}) = L^1_w(m_{V_{p-}})$, it follows that $f \in L^1(m_{V_{p-}})$ if and only if

$$\int_0^1 |f| \, d\left|\langle m_{V_{p-}}, \varphi \rangle\right| = \int_0^1 |f(t)| \left|\langle g_{p-}(t), \varphi \rangle\right| d\lambda(t) < \infty,$$

for every $\varphi \in (L^{p-})^*$. This is what was to be proved.

(iv) $\Rightarrow$ (iii) Let $F_f \in L^{p-}$. Choose an arbitrary $\varphi \in (L^{p-})^*$, i.e., $\varphi \in L^{s_k}$ for at least one $k \in \mathbb{N}$ where $\frac{1}{s_k} + \frac{1}{r_k} = 1$ and $1 \leqslant r_k \uparrow_k p$. Then,

$$\int_0^1 |f(t)| \left|\langle g_{p-}(t), \varphi \rangle\right| d\lambda(t)$$
$$= \int_0^1 |f(t)| \left|\int_0^1 g_{p-}(t)(w) \, \varphi(w) \, d\lambda(w)\right| d\lambda(t)$$
$$\leqslant \int_0^1 |f(t)| \left(\int_0^1 \chi_{[t,1]}(w) \, |\varphi(w)| \, d\lambda(w)\right) d\lambda(t)$$
$$= \int_0^1 |f(t)| \left(\int_0^1 \chi_{[0,w]}(t) \, |\varphi(w)| \, d\lambda(w)\right) d\lambda(t)$$
$$= \int_0^1 \left(\int_0^1 \chi_{[0,w]}(t) \, |f(t)| \, d\lambda(t)\right) |\varphi(w)| \, d\lambda(w)$$
$$= \int_0^1 \left(\int_0^w |f(t)| \, d\lambda(t)\right) |\varphi(w)| \, d\lambda(w)$$
$$= \int_0^1 F_f(w) \, |\varphi(w)| \, d\lambda(w) = \langle F_f, |\varphi| \rangle \leqslant q_k(F_f) \|\varphi\|_{s_k} < \infty.$$

107

(iii) $\Rightarrow$ (iv) Fix $\varphi \in \left(L^{p-}\right)^*$. Then,

$$\left| \int_0^1 |f(t)| \, \langle g_{p-}(t), \varphi \rangle \, d\lambda(t) \right| \leqslant \int_0^1 |f(t)| \, |\langle g_{p-}(t), \varphi \rangle| \, d\lambda(t) < \infty.$$

But,

$$
\begin{aligned}
|\langle F_f, \varphi \rangle| &= \left| \int_0^1 F_f(w) \, \varphi(w) \, d\lambda(w) \right| \\
&\leqslant \int_0^1 \left(\int_0^w |f(t)| \, d\lambda(t) \right) |\varphi(w)| \, d\lambda(w) \\
&= \int_0^1 \left(\int_0^1 \chi_{[0,w]}(t) \, |f(t)| \, d\lambda(t) \right) |\varphi(w)| \, d\lambda(w) \\
&= \int_0^1 \left(\int_0^1 \chi_{[t,1]}(w) \, |f(t)| \, d\lambda(t) \right) |\varphi(w)| \, d\lambda(w) \\
&= \int_0^1 |f(t)| \left(\int_0^1 \chi_{[t,1]}(w) \, |\varphi(w)| \, d\lambda(w) \right) d\lambda(t) \\
&= \int_0^1 |f(t)| \, \langle g_{p-}(t), |\varphi| \rangle \, d\lambda(t) < \infty
\end{aligned}
$$

(by hypothesis) as $\varphi \in L^{s_k}$ for some k implies that also $|\varphi| \in L^{s_k} \subseteq \left(L^{p-}\right)^*$. So, we have shown that $|\langle F_f, \varphi \rangle| < \infty$, for all $\varphi \in \left(L^{p-}\right)^*$, meaning that $F_f \in L^{p-}$. $\quad\square$

Lemma 4.2.3 states some criteria that allow us to decide whether the inclusion $L^1 \subseteq L^1(m_{V_{p-}})$ is proper or not.

Proposition 4.2.5

For each $p \in (1, \infty)$ the inclusion $L^1 \subseteq L^1(m_{V_{p-}})$ is proper.

Proof:

Consider the function $f : [0, 1) \to \mathbb{C}$ defined by

$$f(w) := \frac{1}{(1-w)}, \quad \text{for } w \in [0, 1),$$

which is evidently not an element of L^1. The claim is that the function $F_f : [0, 1) \to \mathbb{C}$ defined by

$$F_f(w) := \int_0^w \frac{1}{1-t} \, d\lambda(t) = -\ln(1-w)$$

is an element of L^{p-}. To see this, fix an arbitrary $k \in \mathbb{N}$. Then, by substituting

$v := 1 - w$, we obtain

$$\int_0^1 \left| F_f(w) \right|^{r_k} d\lambda(w) = \int_0^1 \left(-\ln(1-w) \right)^{r_k} d\lambda(w)$$

$$= \int_0^1 \left(-\ln(v) \right)^{r_k} d\lambda(v)$$

$$= \int_0^1 \tfrac{1}{v^\alpha} \, v^\alpha \left(-\ln(v) \right)^{r_k} d\lambda(v)$$

where $v^{-\alpha} \in L^1$ whenever $0 \leqslant \alpha < 1$. Now, fix an arbitrary $\alpha \in (0,1)$. Then it suffices to show that $v^\alpha \left(-\ln(v) \right)^{r_k} \in L^\infty$. Applying L'Hôpital's rule l times ($l \in \mathbb{N}$) we obtain

$$\lim_{v \to 0} v^\alpha \left(-\ln(v) \right)^{r_k} = \lim_{v \to 0} \frac{\left(-\ln(v) \right)^{r_k}}{v^{-\alpha}}$$

$$= \lim_{v \to 0} \frac{r_k \left(-\ln(v) \right)^{r_k - 1}}{\alpha \, v^{-\alpha}} = \dots$$

$$= \lim_{v \to 0} \frac{r_k(r_k - 1) \cdots (r_k - (l-1)) \left(-\ln(v) \right)^{r_k - l}}{\alpha^l \, v^{-\alpha}}.$$

Continue until $(r_k - l) < 0$. Then,

$$\lim_{v \to 0} \frac{r_k(r_k - 1) \cdots (r_k - (l-1)) (-\ln(v))^{r_k - l}}{\alpha^l \, v^{-\alpha}} = 0,$$

and, hence, $v^\alpha \left(-\ln(v) \right)^{r_k} \in L^\infty$ as required. As $k \in \mathbb{N}$ was chosen arbitrarily, $-\ln(1-w) \in L^{p-}$. Lemma 4.2.3 now implies that $f \in L^1(m_{V_{p-}})$. $\square$

Thus, we obtain the chain of proper inclusions

$$L^{p-} \subsetneqq L^1 \subsetneqq L^1(m_{V_{p-}}).$$

There are still some further connections between various spaces of integrable functions. According to (2.34) we know that

$$\bigcap_{k \in \mathbb{N}} L^1 \left(|(m_{V_{p-}})_k| \right) \subseteq L^1(m_{V_{p-}}) \subseteq \bigcap_{k \in \mathbb{N}} L^1 \left((m_{V_{p-}})_k \right)$$

with all inclusions being continuous. The question arises whether both inclusions involved are proper. The answer for the right-hand inclusion is no. Actually, it is an equality.

Proposition 4.2.6

For each $p \in (1, \infty)$ we have $L^1(m_{V_{p-}}) = \bigcap_{k \in \mathbb{N}} L^1\big((m_{V_{p-}})_k\big)$.

Proof:

It suffices to show that $\bigcap_{k \in \mathbb{N}} L^1\big((m_{V_{p-}})_k\big) \subseteq L^1(m_{V_{p-}})$. Recall, for each $k \in \mathbb{N}$, that the vector measure $(m_{V_{p-}})_k : \mathcal{B}([0,1]) \to L^{r_k}$ is given by

$$(m_{V_{p-}})_k(A) \overset{(4.27)}{=} \int_A g_{r_k}(t)\, d\lambda(t), \quad \text{for } A \in \mathcal{B}([0,1]),$$

where $g_{r_k} : [0,1] \to L^{r_k}$ is defined by $g_{r_k}(t) := \chi_{[t,1]}$. Fix $k \in \mathbb{N}$. In Lemma 5.3 of [28] it was shown that,

$$f \in L^1\big((m_{V_{p-}})_k\big) \quad \Leftrightarrow \quad \int_0^1 |f(t)|\,|\langle g_{r_k}(t), \varphi \rangle|\, d\lambda(t) < \infty, \text{ for all } \varphi \in L^{s_k}, \quad (4.30)$$

where s_k is the conjugate exponent of r_k.

Now, let $f \in \bigcap_{k \in \mathbb{N}} L^1\big((m_{V_{p-}})_k\big)$. Choose an arbitrary $\varphi \in \big(L^{p-}\big)^*$. Then $\varphi \in L^{s_k}$ for some $k \in \mathbb{N}$. It follows from (4.30) that

$$\int_0^1 |f(t)|\,|\langle g_{r_k}(t), \varphi \rangle|\, d\lambda(t) < \infty.$$

Since $g_{p-}(t) = g_{r_k}(t)|_{L^{p-}}$, for each $t \in [0,1]$, it follows that

$$\int_0^1 |f(t)|\,|\langle g_{p-}(t), \varphi \rangle|\, d\lambda(t) < \infty.$$

As φ was chosen arbitrarily this is true for all $\varphi \in \big(L^{p-}\big)^*$ and so Lemma 4.2.3 implies that $f \in L^1(m_{V_{p-}})$. $\square$

What about the inclusion

$$\bigcap_{k \in \mathbb{N}} L^1\big(|(m_{V_{p-}})_k|\big) \subseteq L^1(m_{V_{p-}})? \quad (4.31)$$

In [28, Section 4] it was shown that in the case of the Volterra operator V_r being defined on the space L^r, for $1 \leqslant r \leqslant \infty$, the inclusion becomes an equality whenever $r = 1$ or $r = \infty$, i.e.,

$$L^1(|m_{V_1}|) = L^1(m_{V_1}) \quad \text{resp.} \quad L^1(|m_{V_\infty}|) = L^1(m_{V_\infty}).$$

However, if $1 < r < \infty$, then

$$L^1(|m_{V_r}|) \subsetneq L^1(m_{V_r}).$$

Indeed, the function $f_r : [0, 1) \to \mathbb{C}$ defined by

$$f_r(t) := \frac{(1 - \ln(1 - t)) - r}{r(1 - t)^{1 + (1/r)}(1 - \ln(1 - t))^2} \cdot \chi_{(c,1)},$$

where $c := 1 - \exp(1 - r)$, belongs to $L^1(m_{V_r})$ but not to $L^1(|m_{V_r}|)$, [26, pp. 125–127]. The attempt to construct a similar function, which could verify that the inclusion (4.31) is proper as well, ended without success. So, the question whether the inclusion is indeed proper or reduces to an equality is still open.

A last problem that is worth to be solved concerns the properties of the integration operator

$$I_{m_{V_{p-}}} : L^1(m_{V_{p-}}) \to L^{p-},$$

more precisely, the compactness of $I_{m_{V_{p-}}}$. Recall, for X, Y being Fréchet spaces, that a continuous linear map $T : X \to Y$ is called compact if there is a neighbourhood U of zero in X such that the closure of its range $\overline{T(U)}$ is compact in Y. Since this definition may not always be the best one to check compactness of a given integration operator $I_m : L^1(m) \to X$ associated with a Fréchet-space-valued vector measure $m : \Sigma \to X$, [24] provides alternative characterizations of compactness of I_m. In the case of $I_{m_{V_{p-}}}$, the following theorem turns out to be quite helpful, [24, p. 211 & p. 220].

Theorem 4.2.1
Let X be a Fréchet space and $m : \Sigma \to X$ be a vector measure. The integration operator $I_m : L^1(m) \to X$ is compact if and only if there exists an index $l \in \mathbb{N}$ such that $I_{m_k} : L^1(m_k) \to X_k$ is compact and $L^1(m_k) = L^1(m_l)$, for every $k \geqslant l$. $\quad\square$

Having this theorem and the results established in [28] available the question concerning the compactness of the integration operator $I_{m_{V_{p-}}}$ is answered immediately. To show that $I_{m_{V_{p-}}} : L^1(m_{V_{p-}}) \to L^{p-}$ is compact we needed to prove the existence of an index $l \in \mathbb{N}$ such that $I_{m_{V_{r_k}}} : L^1(m_{V_{r_k}}) \to L^{r_k}$ is compact, for all $k \geqslant l$. However, the investigations in [28, Proposition 5.5] revealed that *none* of the integration operators $I_{m_{V_r}} : L^1(m_{V_r}) \to L^r$, for $1 < r < \infty$, is compact. Thus, the requirements for the compactness of $I_{m_{V_{p-}}}$ as stated in Theorem 4.2.1 are not met and we can draw the following conclusion.

Proposition 4.2.7

For each $p \in (1, \infty)$ the integration operator $I_{m_{V_{p-}}} : L^1(m_{V_{p-}}) \to L^{p-}$ is not compact.

□

4.3 The convolution operator

In this section we turn to integration on a certain class of topological groups. Let $(G, +)$ be an (additive) compact Abelian group and consider the finite measure space $(G, \mathcal{B}(G), \mu)$ where μ is normalized Haar measure on G, i.e., $\mu(G) = 1$, and $\mathcal{B}(G)$ is the Borel σ-algebra of G. For $p \in (1, \infty)$ fixed, let $L^{p-}(G)$ be the Fréchet function space as defined in Example 2.3.1, this time defined on G instead of the interval $[0, 1]$. That is, $L^{p-}(G) = \bigcap_{k \in \mathbb{N}} L^{r_k}(G)$ with $1 \leqslant r_k \uparrow_k p$, equipped with the norms

$$q_k(f) := \left(\int_G |f|^{r_k} \, d\mu \right)^{1/r_k} = \|f\|_{r_k}, \quad \text{for } f \in L^{p-}(G),$$

for $k \in \mathbb{N}$.

For $g \in L^1(G)$ fixed, define on $L^{p-}(G)$ the *convolution operator* $C_g^{p-} : L^{p-}(G) \to L^{p-}(G)$ given by $f \mapsto C_g^{p-}(f)$ where

$$C_g^{p-}(f)(x) = (f * g)(x) := \int_G f(y) \, g(x - y) \, d\mu(y), \quad \text{for } \mu\text{-almost every } x \in G.$$

(4.32)

Since for functions $f \in L^{r_k}(G)$, with $k \in \mathbb{N}$, and $g \in L^1(G)$ the resulting function $f * g$ is an element of $L^{r_k}(G)$ (see p. 50), it is clear that indeed $C_g^{p-}(f) \in L^{p-}(G)$, for all $f \in L^{p-}(G)$. Note that the convolution operator is linear. It is continuous as well as seen via the next result.

Proposition 4.3.1

For each $1 < p < \infty$ and $g \in L^1(G)$, the convolution operator $C_g^{p-} : L^{p-}(G) \to L^{p-}(G)$ is continuous.

Proof:

Let $f \in L^{p-}(G)$. Then $f \in L^{r_k}(G)$, for all $k \in \mathbb{N}$, and since g is an element of $L^1(G)$, we obtain by applying (2.38) that

$$q_k\big(C_g^{p-}(f)\big) = q_k\big(f * g\big) \overset{(2.38)}{\leqslant} q_k(f) \underbrace{\|g\|_1}_{=:M} = M \, q_k(f),$$

for all $k \in \mathbb{N}$. Thus, C_g^{p-} is continuous. □

We now take a closer look at the vector measure $m_{C_g^{p^-}}$ associated with the operator $C_g^{p^-}$, i.e., the vector measure $m_{C_g^{p^-}} : \mathcal{B}(G) \to L^{p^-}(G)$ defined by

$$m_{C_g^{p^-}}(A) := C_g^{p^-}(\chi_A) = \chi_A * g, \quad \text{for } A \in \mathcal{B}(G).$$

Clearly, $m_{C_g^{p^-}}$ is finitely additive. To see that $m_{C_g^{p^-}}$ is a vector measure let $\{A_j\}_{j \in \mathbb{N}} \subseteq \mathcal{B}(G)$ be any sequence satisfying $A_j \downarrow_j \varnothing$. For $k \in \mathbb{N}$ fixed, we obtain that

$$q_k\big(m_{C_g^{p^-}}(A_j)\big) = q_k\big(\chi_{A_j} * g\big) \overset{(2.38)}{\leqslant} q_k\big(\chi_{A_j}\big)\,\|g\|_1$$
$$= \left(\int_G |\chi_{A_j}|^{r_k}\, d\mu\right)^{1/r_k}\|g\|_1$$
$$= \mu(A_j)^{1/r_k}\|g\|_1,$$

for all $j \in \mathbb{N}$. But, μ being a measure, it follows that

$$0 \leqslant \lim_{j\to\infty} q_k\big(m_{C_g^{p^-}}(A_j)\big) \leqslant \lim_{j\to\infty} \mu(A_j)^{1/r_k}\|g\|_1 = 0$$

implying that $m_{C_g^{p^-}}$ is σ-additive.

For each $g \in L^1(G)$ define the reflected function $\tilde{g} \in L^1(G)$ by $\tilde{g}(x) := g(-x)$, for all $x \in G$. Then we obtain, for each $A \in \mathcal{B}(G)$, that

$$C_g^{p^-}(\chi_A)(x) = \int_G \chi_A(y)\, g(x-y)\, d\mu(y) = \int_A g(x-y)\, d\mu(y) = \int_A \tilde{g}(y-x)\, d\mu(y),$$

for $x \in G$. For each $\varphi \in \big(L^{p^-}(G)\big)^* = \bigcup_{k \in \mathbb{N}} L^{s_k}(G)$ with $\frac{1}{r_k} + \frac{1}{s_k} = 1$ (see Example 2.3.1 (i) with G in place of $[0,1]$), we can deduce a formula for the scalar measure $\langle m_{C_g^{p^-}}, \varphi \rangle : \mathcal{B}(G) \to \mathbb{C}$ given by

$$\langle m_{C_g^{p^-}}, \varphi \rangle(A) := \langle m_{C_g^{p^-}}(A), \varphi \rangle, \quad \text{for } A \in \mathcal{B}(G).$$

Namely, by applying Fubini's Theorem 2.2.3, the expression can also be written as

$$\langle m_{C_g^{p^-}}(A), \varphi \rangle = \langle C_g^{p^-}(\chi_A), \varphi \rangle$$
$$= \int_G C_g^{p^-}(\chi_A)(x)\, \varphi(x)\, d\mu(x)$$
$$= \int_G (\chi_A * g)(x)\, \varphi(x)\, d\mu(x)$$
$$= \int_G \left(\int_G \chi_A(y)\, g(x-y)\, d\mu(y)\right) \varphi(x)\, d\mu(x)$$
$$= \int_G \chi_A(y) \left(\int_G g(x-y)\, \varphi(x)\, d\mu(x)\right) d\mu(y)$$

$$= \int_A \left(\int_G \tilde{g}(y-x)\, \varphi(x)\, d\mu(x) \right) d\mu(y)$$

$$= \int_A (\tilde{g} * \varphi)\, d\mu, \tag{4.33}$$

for each $A \in \mathcal{B}(G)$. Note that $\tilde{g} * \varphi \in L^q(G)$ for at least one $q \in (1, \infty)$. This is due to the fact that $\tilde{g} \in L^1(G)$ and $\varphi \in \left(L^{p-}(G) \right)^* = \bigcup_{k \in \mathbb{N}} L^{s_k}(G)$ implying that $\varphi \in L^q(G)$ for at least one $q \in (1, \infty)$.

Our main investigations will concentrate on the space $L^1(m_{C_g^{p-}})$. Let the measurable function $f : G \to \mathbb{C}$ be $m_{C_g^{p-}}$-integrable, i.e.,

$$\int_G |f|\, d|\langle m_{C_g^{p-}}, \varphi \rangle| < \infty, \quad \text{for all } \varphi \in \left(L^{p-}(G) \right)^*,$$

and, for each $A \in \mathcal{B}(G)$, there exists an element $\int_A f\, dm_{C_g^{p-}} \in L^{p-}(G)$ satisfying

$$\left\langle \int_A f\, dm_{C_g^{p-}}, \varphi \right\rangle = \int_A f\, d\langle m_{C_g^{p-}}, \varphi \rangle, \quad \text{for } \varphi \in \left(L^{p-}(G) \right)^*.$$

Then, for each $A \in \mathcal{B}(G)$, we have via Fubini's Theorem 2.2.3 that

$$\int_A f\, d\langle m_{C_g^{p-}}, \varphi \rangle \overset{(4.33)}{=} \int_A f(y)\, (\tilde{g} * \varphi)(y)\, d\mu(y)$$

$$= \int_A f(y) \left(\int_G \tilde{g}(y-x)\, \varphi(x)\, d\mu(x) \right) d\mu(y)$$

$$= \int_G (f\chi_A)(y) \left(\int_G g(x-y)\, \varphi(x)\, d\mu(x) \right) d\mu(y)$$

$$= \int_G \left(\int_G (f\chi_A)(y)\, g(x-y)\, d\mu(y) \right) \varphi(x)\, d\mu(x)$$

$$= \int_G ((f\chi_A) * g)(x)\, \varphi(x)\, d\mu(x)$$

$$= \langle (f\chi_A) * g, \varphi \rangle,$$

where the last equality is only possible if $(f\chi_A) * g \in L^{p-}(G)$. Hence, the indefinite integral of $f \in L^1(m_{C_g^{p-}})$ over $A \in \mathcal{B}(G)$ is given by

$$\int_A f\, dm_{C_g^{p-}} = (f\chi_A) * g, \quad \text{for } A \in \mathcal{B}(G), \tag{4.34}$$

provided that $(f\chi_A) * g \in L^{p-}(G)$, for every $A \in \mathcal{B}(G)$. In this case, the vector measure $m_{C_g^{p-}, f} : \mathcal{B}(G) \to L^{p-}(G)$ associated with the indefinite integral of f is given by

$$m_{C_g^{p^-},f}(A) := \int_A f \, dm_{C_g^{p^-}} = (f\chi_A) * g, \quad \text{for } A \in \mathcal{B}(G). \tag{4.35}$$

In order to apply the theory of Chapter 3 we need to show that $C_g^{p^-}$ is μ-determined. To prove this and some of the forthcoming assertions we will make use of the fact that for each $A \in \mathcal{B}(G)$,

$$m_{C_g^{p^-}}(A) \in L^{p^-}(G) = \bigcap_{k \in \mathbb{N}} L^{r_k}(G) \subseteq L^{r_k}(G), \quad \text{for all } k \in \mathbb{N}.$$

Hence, it makes sense to take the local-Banach-space-valued vector measures

$$(m_{C_g^{p^-}})_k : \Pi_k \circ m_{C_g^{p^-}} : \mathcal{B}(G) \to L^{r_k}(G), \quad \text{for } k \in \mathbb{N},$$

into account. For fixed $k \in \mathbb{N}$, we thereby have

$$m_{C_g^{p^-}}(A) = (m_{C_g^{p^-}})_k(A) = \chi_A * g \in L^{r_k}(G),$$

for all $A \in \mathcal{B}(G)$. Note, for each $k \in \mathbb{N}$, that the vector measure $(m_{C_g^{p^-}})_k$ coincides with the vector measure $m_{C_g^{r_k}} : \mathcal{B}(G) \to L^{r_k}(G)$ as investigated in [25]. We follow that notation and write

$$(m_{C_g^{p^-}})_k(A) =: m_{C_g^{r_k}}(A) = C_g^{r_k}(\chi_A), \quad \text{for } A \in \mathcal{B}(G), \tag{4.36}$$

where $C_g^{r_k} : L^{r_k}(G) \to L^{r_k}(G)$ is the Banach-space-valued operator of convolution with $g \in L^1(G)$.

Proposition 4.3.2

For each $1 < p < \infty$ and $g \in L^1(G)\backslash\{0\}$, the convolution operator $C_g^{p^-} : L^{p^-}(G) \to L^{p^-}(G)$ is μ-determined.

Proof:

We apply Lemma 3.2.2 which states that the operator $C_g^{p^-}$ is μ-determined if and only if the $m_{C_g^{p^-}}$-null sets and the μ-null sets coincide. We show at first that $\mathcal{N}_0(m_{C_g^{p^-}}) \subseteq \mathcal{N}_0(\mu)$ holds. Let $A \in \mathcal{N}_0(m_{C_g^{p^-}})$, that is,

$$m_{C_g^{p^-}}(B) = \chi_B * g = 0 \in L^{p^-}(G),$$

for all $B \in \mathcal{B}(G)$ with $B \subseteq A$. Since $L^{p^-}(G) \subseteq L^{r_1}(G)$, the discussion prior to this proposition gives that

$$m_{C_g^{r_1}}(B) = \chi_B * g = 0 \in L^{r_1}(G),$$

115

for all $B \in \mathcal{B}(G)$ with $B \subseteq A$, meaning that $A \in \mathcal{N}_0(m_{C_g^{r_1}})$. But, by applying Lemma 2.2 of [25] to the Banach-space-valued vector measure $m_{C_g^{r_1}}$, we obtain that $A \in \mathcal{N}_0(\mu)$. Hence, $\mathcal{N}_0(m_{C_g^{p-}}) \subseteq \mathcal{N}_0(\mu)$.

The inclusion $\mathcal{N}_0(\mu) \subseteq \mathcal{N}_0(m_{C_g^{p-}})$ is an immediate consequence of Lemma 3.2.1. Let $A \in \mathcal{B}(G)$ be any μ-null set. Then χ_B is a μ-null function, for all $B \in \mathcal{B}(G)$ with $B \subseteq A$ and, thus, by Lemma 3.2.1 also an $m_{C_g^{p-}}$-null function. But this means that A is an $m_{C_g^{p-}}$-null set, i.e., $\mathcal{N}_0(\mu) \subseteq \mathcal{N}_0(m_{C_g^{p-}})$. Thus, the μ-null sets and the $m_{C_g^{p-}}$-null sets coincide and so, by Lemma 3.2.2, C_g^{p-} is μ-determined. $\quad\square$

According to the previous result, the μ-determinedness of C_g^{p-} implies that the μ-null functions and the $m_{C_g^{p-}}$-null functions coincide, i.e., that

$$\mathcal{N}(\mu) = \mathcal{N}(m_{C_g^{p-}}),$$

and the same is true for the null sets, i.e.,

$$\mathcal{N}_0(\mu) = \mathcal{N}_0(m_{C_g^{p-}}).$$

By applying the theory of Chapter 3 we can derive some additional facts about $m_{C_g^{p-}}$ and $L^1(m_{C_g^{p-}})$. Proposition 3.2.2 yields that each $f \in L^{p-}(G)$ is $m_{C_g^{p-}}$-integrable, i.e.,

$$L^{p-}(G) \subseteq L^1(m_{C_g^{p-}}),$$

and so, for each $f \in L^{p-}(G)$, the equation

$$C_g^{p-}(f\chi_A) = (f\chi_A) * g \overset{(4.34)}{=} \int_A f \, dm_{C_g^{p-}}, \quad \text{for } A \in \mathcal{B}(G),$$

holds. Furthermore, Theorem 3.3.1 ensures that $L^1(m_{C_g^{p-}})$ is the optimal domain of C_g^{p-} and its optimal extension is the integration operator $I_{m_{C_g^{p-}}} : L^1(m_{C_g^{p-}}) \to L^{p-}(G)$ given by

$$I_{m_{C_g^{p-}}}(f) := \int_G f \, dm_{C_g^{p-}} = f * g, \quad \text{for } f \in L^1(m_{C_g^{p-}}).$$

Moreover, since $L^{p-}(G)$ is reflexive, $L^1(m_{C_g^{p-}}) = L_w^1(m_{C_g^{p-}})$.

This allows us to study further the space $L^1(m_{C_g^{p-}})$. Let us first establish the following two statements.

Lemma 4.3.1

Let $1 < p < \infty$ and $g \in L^1(G)$. Then, $L^1(m_{C_g^{p-}}) \subseteq L^1(G)$.

Proof:

Let f be an $m_{C_g^{p-}}$-integrable function. Since $L^{p-}(G)$ is reflexive, $f \in L^1(m_{C_g^{p-}})$ is equivalent to $f \in L^1_w(m_{C_g^{p-}})$ meaning that

$$\int_G |f|\,d\big|\langle m_{C_g^{p-}}, \varphi \rangle\big| < \infty, \quad \text{for all } \varphi \in \big(L^{p-}(G)\big)^*,$$

where $\big(L^{p-}(G)\big)^* = \bigcup_{k \in \mathbb{N}} L^{s_k}(G)$ with $\frac{1}{r_k} + \frac{1}{s_k} = 1$ and $1 \leqslant r_k \uparrow_k p$. In particular, for any fixed $k \in \mathbb{N}$ we have

$$\int_G |f|\,d\big|\langle m_{C_g^{p-}}, \varphi \rangle\big| < \infty, \quad \text{for all } \varphi \in L^{s_k}(G). \tag{4.37}$$

On the other hand, as discussed prior to Proposition 4.3.2,

$$m_{C_g^{p-}}(A) = m_{C_g^{r_k}}(A) = \chi_A * g \in L^{p-}(G) \subseteq L^{r_k}(G),$$

for all $A \in \mathcal{B}(G)$, and thus,

$$\big\langle m_{C_g^{p-}}, \varphi \big\rangle = \big\langle m_{C_g^{r_k}}, \varphi \big\rangle$$

as scalar measures. Hence, (4.37) shows that

$$\int_G |f|\,d\big|\langle m_{C_g^{r_k}}, \varphi \rangle\big| < \infty, \quad \text{for all } \varphi \in L^{s_k}(G) = \big(L^{r_k}(G)\big)^*,$$

meaning that $f \in L^1_w(m_{C_g^{r_k}})$ and since $L^{r_k}(G)$ is reflexive, $f \in L^1(m_{C_g^{r_k}})$. By Theorem 1.1 (v) in [25] it is known that $L^1(m_{C_g^{r_k}}) \subseteq L^1(G)$. Hence, $f \in L^1(G)$ and the assertion of this lemma holds. $\square$

Lemma 4.3.2
Let $1 < p < \infty$, $g \in L^1(G)$ and $f \in L^1(m_{C_g^{p-}})$. Then $(f\chi_A) * g \in L^{p-}(G)$, for all $A \in \mathcal{B}(G)$.

Proof:

Let $f \in L^1(m_{C_g^{p-}})$. In the proof of Lemma 4.3.1 it was shown that necessarily $f \in L^1(m_{C_g^{r_k}})$, for all $k \in \mathbb{N}$. For any fixed $k \in \mathbb{N}$, Theorem 1.1 (vi) of [25] implies that $(f\chi_A) * g \in L^{r_k}(G)$. But, as $k \in \mathbb{N}$ was chosen arbitrarily it follows that $(f\chi_A) * g \in L^{r_k}(G)$, for all $k \in \mathbb{N}$, and consequently $(f\chi_A) * g \in L^{p-}(G)$. $\square$

The previous thoughts give an idea for characterizing the space $L^1(m_{C_g^{p-}})$. Indeed, whenever $g \geqslant 0$, we obtain the following result.

Proposition 4.3.3

Let $1 < p < \infty$ and let $g \in L^1(G)\backslash\{0\}$ satisfy $g \geqslant 0$. Then,

$$L^1(m_{C_g^{p-}}) = \{f \in L^1(G) : (f\chi_A) * g \in L^{p-}(G), \text{ for all } A \in \mathcal{B}(G)\}.$$

Proof:

Let $f \in L^1(m_{C_g^{p-}})$. Then $f \in L^1(G)$ by Lemma 4.3.1. Moreover, Lemma 4.3.2 yields that $(f\chi_A) * g \in L^{p-}(G)$, for all $A \in \mathcal{B}(G)$. Hence,

$$L^1(m_{C_g^{p-}}) \subseteq \{f \in L^1(G) : (f\chi_A) * g \in L^{p-}(G), \text{ for all } A \in \mathcal{B}(G)\}.$$

Conversely, let $f \in L^1(G)$ satisfy $(f\chi_A) * g \in L^{p-}(G)$, for all $A \in \mathcal{B}(G)$. Fix $k \in \mathbb{N}$. Since $L^{p-}(G) \subseteq L^{r_k}(G)$ it follows that

$$(f\chi_A) * g \in L^{r_k}(G), \quad \text{for all } A \in \mathcal{B}(G).$$

By Proposition 3.2 of [25] we can conclude that $f \in L^1(m_{C_g^{r_k}})$ for the Banach-space-valued vector measure $m_{C_g^{r_k}} : \mathcal{B}(G) \to L^{r_k}(G)$. Then $f \in L_w^1(m_{C_g^{r_k}})$ and so,

$$\int_G |f| \, d|\langle m_{C_g^{r_k}}, \varphi \rangle| < \infty, \quad \text{for all } \varphi \in \left(L^{r_k}(G)\right)^* = L^{s_k}(G).$$

However, as noted prior to Proposition 4.3.2,

$$m_{C_g^{p-}}(A) = m_{C_g^{r_k}}(A) = C_g^{r_k}(\chi_A) = \chi_A * g, \quad \text{for } A \in \mathcal{B}(G).$$

Hence,

$$\int_G |f| \, d|\langle m_{C_g^{p-}}, \varphi \rangle| < \infty, \quad \text{for all } \varphi \in L^{s_k}(G). \tag{4.38}$$

But, $k \in \mathbb{N}$ was arbitrary, so (4.38) holds for all $k \in \mathbb{N}$ and since $\left(L^{p-}(G)\right)^* = \bigcup_{k \in \mathbb{N}} L^{s_k}(G)$, it follows that

$$\int_G |f| \, d|\langle m_{C_g^{p-}}, \varphi \rangle| < \infty, \quad \text{for all } \varphi \in \left(L^{p-}(G)\right)^*,$$

that is, $f \in L_w^1(m_{C_g^{p-}})$ and since $L^{p-}(G)$ is reflexive, $f \in L^1(m_{C_g^{p-}})$. $\square$

Remark 4.3.1

Observe that the first part of the proof of Proposition 4.3.3 did not use the condition $g \geqslant 0$ and so the inclusion

$$L^1(m_{C_g^{p-}}) \subseteq \{f \in L^1(G) : (f\chi_A) * g \in L^{p-}(G), \text{ for all } A \in \mathcal{B}(G)\}$$

holds for any $\mathbb{C}$-valued function $g \in L^1(G)$.

In contrast to the investigations above, we now change the conditions on g. From now on g will always be an element of the smaller space $L^{p-}(G) \subseteq L^1(G)$. Proposition 4.3.3 gave a characterization of the space $L^1(m_{C_g^{p-}})$ for $g \in L^1(G)$ satisfying $g \geqslant 0$; the question now is whether a characterization of $L^1(m_{C_g^{p-}})$ is also possible for $g \in L^{p-}(G)$, but without assuming $g \geqslant 0$?

So, let $g \in L^{p-}(G)$. Define, for $y \in G$ fixed, the *translation operator* $\tau_y : L^{p-}(G) \to L^{p-}(G)$ by

$$\tau_y(g) := g(\cdot - y). \tag{4.39}$$

The fact that the Haar measure μ is translation invariant ensures that the translation operator τ_y is continuous on $L^{p-}(G)$. To see this, choose an arbitrary function $g \in L^{p-}(G)$. Then, for each $k \in \mathbb{N}$, we obtain that

$$\begin{aligned} q_k(\tau_y(g)) &= \left(\int_G |\tau_y(g)|^{r_k} \, d\mu \right)^{1/r_k} \\ &= \left(\int_G |g(x - y)|^{r_k} \, d\mu(x) \right)^{1/r_k} \\ &= \left(\int_G |g(x)|^{r_k} \, d\mu(x) \right)^{1/r_k} = q_k(g). \end{aligned} \tag{4.40}$$

This shows that $\tau_y(g) \in L^{p-}(G)$, for each $y \in G$, and that the operators $\{\tau_y \,|\, y \in G\}$ are all continuous on $L^{p-}(G)$.

Fix $g \in L^{p-}(G)$. Associated with τ_y define the $L^{p-}(G)$-valued function $F_g^{p-} : G \to L^{p-}(G)$ by

$$F_g^{p-}(y) := \tau_y(g) = g(\cdot - y), \quad \text{for } y \in G. \tag{4.41}$$

The function F_g^{p-} will serve us well in the following investigations. This is due to its special properties, two of which we state here.

Proposition 4.3.4

For each $1 < p < \infty$ and $g \in L^{p-}(G)$, the function F_g^{p-} as defined in (4.41) is continuous.

Proof:

Fix $k \in \mathbb{N}$ and note that $g \in L^{r_k}(G)$. Hence, by the uniform continuity of the function $F_g^{r_k} : G \to L^{r_k}(G)$, [30, p. 3], for any given $\varepsilon > 0$ there exists a neighbourhood V_k of zero in G such that

$$q_k\big(F_g^{p-}(w) - F_g^{p-}(y)\big) = q_k\big(\tau_w(g) - \tau_y(g)\big) < \varepsilon$$

whenever $w - y \in V_k$. $\quad\square$

Proposition 4.3.5

For each $1 < p < \infty$ and $g \in L^{p-}(G)$, the function F_g^{p-} as defined in (4.41) is Bochner μ-integrable.

Proof:

Let $g \in L^{p-}(G)$. Since F_g^{p-} is continuous and G is compact, $F_g^{p-}(G)$ is compact in the Fréchet space $L^{p-}(G)$ meaning that $F_g^{p-}(G)$ is a compact metric space and thus, is separable, [22, pp. 18–19]. Hence, $F_g^{p-}(G)$ is a Suslin space and it follows from [34, pp. 67–68] that F_g^{p-} is strongly μ-measurable.

Moreover, we have for each $k \in \mathbb{N}$

$$q_k\big(F_g^{p-}(y)\big) \stackrel{(4.41)}{=} q_k\big(\tau_y(g)\big) \stackrel{(4.40)}{=} q_k(g), \quad \text{for all } y \in G, \tag{4.42}$$

and thus,

$$\int_G q_k\big(F_g^{p-}(y)\big)\, d\mu \stackrel{(4.42)}{=} \int_G q_k(g)\, d\mu = q_k(g)\, \mu(G) < \infty.$$

Hence, F_g^{p-} is Bochner μ-integrable. $\quad\square$

For the following proposition and main result of this section we take a second time a closer look at the local-Banach-space-valued vector measures $(m_{C_g^{p-}})_k : \mathcal{B}(G) \to L^{r_k}(G)$, for $k \in \mathbb{N}$, given by

$$m_{C_g^{r_k}}(A) \stackrel{(4.36)}{=} (m_{C_g^{p-}})_k(A) = \int_A g(x - y)\, d\mu(x) = \int_A \tau_y(g)(x)\, d\mu(x)$$

where $\tau_y(g) := g(\cdot - y)$ is again the translation operator, this time having $L^{r_k}(G)$ as domain and codomain. It follows from Lemma 2.3 (ii) of [25], for each $k \in \mathbb{N}$, that the variation measure $|m_{C_g^{r_k}}| = |(m_{C_g^{p-}})_k| : \mathcal{B}(G) \to [0, \infty]$ of $m_{C_g^{r_k}}$ is given by

$$|m_{C_g^{r_k}}|(A) = |(m_{C_g^{p-}})_k|(A) = \int_A \big\|F_g^{r_k}(y)\big\|_{r_k}\, d\mu = \|g\|_{r_k}\mu(A), \quad \text{for } A \in \mathcal{B}(G), \tag{4.43}$$

where $F_g^{r_k}$ is defined as in (4.41) this time, however, considered as an $L^{r_k}(G)$-valued function. Note that each $F_g^{r_k}$ is Bochner μ-integrable as well, [25, p. 532].

Now we can prove the main result of this section; it states six equivalent assertions that hold whenever g is an element of $L^{p^-}(G)\backslash\{0\}$.

Proposition 4.3.6

For $1 < p < \infty$ and a non-zero function $g \in L^1(G)$ the following assertions are equivalent:

(i) $g \in L^{p^-}(G)$.

(ii) $m_{C_g^{p^-}}$ is of finite variation.

(iii) $\bigcap_{k\in\mathbb{N}} L^1\big(|(m_{C_g^{p^-}})_k|\big) = L^1(G)$.

(iv) $L^1(m_{C_g^{p^-}}) = L^1(G)$.

(v) $\bigcap_{k\in\mathbb{N}} L^1\big(|(m_{C_g^{p^-}})_k|\big) = L^1(m_{C_g^{p^-}})$.

(vi) $m_{C_g^{p^-}}$ has an $L^{p^-}(G)$-valued Bochner density $F_g^{p^-}$.

Proof:

(i) $\Leftrightarrow$ (ii) Assume that $g \in L^{p^-}(G)$, meaning that $g \in L^{r_k}(G)$, for all $k \in \mathbb{N}$. According to [25, Theorem 1.2] each of the vector measures $m_{C_g^{r_k}} = (m_{C_g^{p^-}})_k :$ $\mathcal{B}(G) \to L^{r_k}(G)$, where $k \in \mathbb{N}$, has finite variation and thus, $m_{C_g^{p^-}} : \mathcal{B}(G) \to L^{p^-}(G)$ is of finite variation as well.

Conversely, let $m_{C_g^{p^-}}$ be of finite variation. By definition this means that each of the vector measures $(m_{C_g^{p^-}})_k = m_{C_g^{r_k}} : \mathcal{B}(G) \to L^{r_k}(G)$, for $k \in \mathbb{N}$, has finite variation. But, for $k \in \mathbb{N}$ fixed, this is by [25, Theorem 1.2] equivalent to the requirement that $g \in L^{r_k}(G)$. As this is true for all $k \in \mathbb{N}$ it follows that $g \in L^{p^-}(G)$.

(i) $\Rightarrow$ (iii) Let $g \in L^{p^-}(G)$. Then $g \in L^{r_k}(G)$, for all $k \in \mathbb{N}$. Theorem 1.2 of [25] (see (i) $\Leftrightarrow$ (vi)) yields then that

$$\bigcap_{k\in\mathbb{N}} L^1\big(|(m_{C_g^{p^-}})_k|\big) = \bigcap_{k\in\mathbb{N}} L^1\big(|m_{C_g^{r_k}}|\big) = L^1(G).$$

(iii) $\Rightarrow$ (iv) It follows from (2.34) and (iii) that

$$L^1(G) = \bigcap_{k\in\mathbb{N}} L^1\big(|(m_{C_g^{p^-}})_k|\big) \subseteq L^1(m_{C_g^{p^-}}) \tag{4.44}$$

and from Lemma 4.3.1 that

$$L^1(m_{C_g^{p^-}}) \subseteq L^1(G). \tag{4.45}$$

It is then clear that (iv) holds.

121

(iii) $\Rightarrow$ (v) follows immediately from (4.44) and (4.45).

(iv) $\Rightarrow$ (i) Let $L^1(m_{C_g^{p-}}) = L^1(G)$ be true. By Remark 4.3.1 we then know that

$$L^1(G) = L^1(m_{C_g^{p-}}) \subseteq \{f \in L^1(G) : (f\chi_A)*g \in L^{p-}(G), \text{ for all } A \in \mathcal{B}(G)\} \subseteq L^1(G).$$

Consequently, for any $f \in L^1(G)$, the function $f * g$ is an element of $L^{p-}(G) = \bigcap_{k \in \mathbb{N}} L^{r_k}(G)$. So, for $k \in \mathbb{N}$ fixed, $f*g \in L^{r_k}(G)$, for all $f \in L^1(G)$, yields according to [15, Lemma 35.11] that $g \in L^{r_k}(G)$. As this is true for all $k \in \mathbb{N}$ we obtain that $g \in L^{r_k}(G)$, for all $k \in \mathbb{N}$, and thus, $g \in L^{p-}(G)$.

(v) $\Rightarrow$ (i) Assume that

$$\bigcap_{k \in \mathbb{N}} L^1\big(|(m_{C_g^{p-}})_k|\big) = L^1(m_{C_g^{p-}})$$

holds. Making use of Proposition 3.2.2 and the fact that $L^{p-}(G)$ contains the $\mathcal{B}(G)$-simple functions, we obtain the following chain of inclusions:

$$\text{sim}\big(\mathcal{B}(G)\big) \subseteq L^{p-}(G) \subseteq L^1(m_{C_g^{p-}}) = \bigcap_{k \in \mathbb{N}} L^1\big(|(m_{C_g^{p-}})_k|\big).$$

But this means that $\chi_G \in \bigcap_{k \in \mathbb{N}} L^1\big(|(m_{C_g^{p-}})_k|\big)$. Furthermore, since $\rho_k(\chi_G) < \infty$, for all $k \in \mathbb{N}$, the equalities

$$\rho_k(\chi_G) = \int_G |\chi_G| \, d|(m_{C_g^{p-}})_k| \stackrel{(4.43)}{=} \|g\|_{r_k} \underbrace{\mu(G)}_{=1} = q_k(g)$$

imply that $q_k(g) < \infty$, for all $k \in \mathbb{N}$, and hence, $g \in L^{p-}(G)$.

(i) $\Rightarrow$ (vi) It is known by Proposition 4.3.5 that $F_g^{p-} : G \to L^{p-}(G)$ as defined in (4.41) is Bochner μ-integrable. Furthermore, for each $A \in \mathcal{B}(G)$, we obtain that

$$(\text{B}) - \int_A F_g^{p-}(y) \, d\mu(y) = \chi_A * g = m_{C_g^{p-}}(A).$$

This follows from [25, Lemma 2.3 (ii)] applied in each Banach space $L^{r_k}(G)$ to $m_{C_g^{r_k}}$, for $k \in \mathbb{N}$, after noting that F_g^{p-} can be interpreted as being $L^{r_k}(G)$-valued, where it is denoted in [25] by $F_g^{r_k}$.

(vi) $\Rightarrow$ (ii) Let $F_g^{p-} : G \to L^{p-}(G)$, as given in (4.41), be the Bochner density of $m_{C_g^{p-}}$. Fix an arbitrary $k \in \mathbb{N}$ and consider the local-Banach-space-valued vector measure $(m_{C_g^{p-}})_k = m_{C_g^{r_k}} : \mathcal{B}(G) \to L^{r_k}(G)$. Then $F_g^{p-} : G \to L^{p-}(G) \subseteq L^{r_k}(G)$ is the Bochner density of $m_{C_g^{r_k}}$. It follows from [25, Lemma 2.3] that $m_{C_g^{r_k}} = (m_{C_g^{p-}})_k$ has finite variation. As this is true for all $k \in \mathbb{N}$, the vector measure $m_{C_g^{p-}}$ is by definition of finite variation. $\square$

From the results we have obtained in Proposition 4.3.6 we can deduce an additional property of the integration operator

$$I_{m_{C_g^{p-}}} : L^1(m_{C_g^{p-}}) \to L^{p-}(G),$$

namely the compactness of $I_{m_{C_g^{p-}}}$. For the proof we can fall back on Theorem 4.2.1 again.

Proposition 4.3.7

Let $g \in L^1(G)$ be a non-zero function. Then the following assertion is equivalent to the assertions (i)–(vi) of Proposition 4.3.6:

(vii) The integration operator $I_{m_{C_g^{p-}}} : L^1(m_{C_g^{p-}}) \to L^{p-}(G)$ is compact.

Proof:

First, assume (i) of Proposition 4.3.6, i.e., $g \in L^{p-}(G)\backslash\{0\}$. According to the implication (i) $\Rightarrow$ (iv) of Proposition 4.3.6 we have $L^1(m_{C_g^{p-}}) = L^1(G)$. Let $l \in \mathbb{N}$ be the smallest integer such that $q_l(g) > 0$ yielding that $g \in L^{r_l}(G)\backslash\{0\}$. It follows from Theorem 1.2 of [25] and the discussion prior to Proposition 4.3.2 that also $L^1\big((m_{C_g^{p-}})_l\big) = L^1(m_{C_g^{r_l}}) = L^1(G)$ and hence,

$$L^1(m_{C_g^{p-}}) = L^1\big((m_{C_g^{p-}})_l\big).$$

Since the semi-norms $\{q_k\}_{k\in\mathbb{N}}$ are increasing in $L^{p-}(G)$ we have $q_k(g) \geqslant q_l(g) > 0$, for all $k \geqslant l$. The same argument as for l shows that

$$L^1(m_{C_g^{p-}}) = L^1\big((m_{C_g^{p-}})_k\big), \quad \text{for all } k \geqslant l.$$

As for each $k \geqslant l$ we have $g \in L^{r_k}(G)\backslash\{0\}$, Theorem 1.2 of [25] implies that the integration map

$$I_{(m_{C_g^{p-}})_k} : L^1\big((m_{C_g^{p-}})_k\big) \to L^{r_k}(G)$$

is compact, for all $k \geqslant l$. Hence, all assumptions of Theorem 4.2.1 are fulfilled for $m_{C_g^{p-}} : \mathcal{B}(G) \to L^{p-}(G)$ and we can conclude from that result that

$$I_{m_{C_g^{p-}}} : L^1(m_{C_g^{p-}}) \to L^{p-}(G)$$

is compact, i.e., (vii) holds.

Now assume $I_{m_{C_g^{p-}}} : L^1(m_{C_g^{p-}}) \to L^{p-}(G)$ is compact. Choose $l \in \mathbb{N}$ as in

Theorem 4.2.1. By that result

$$I_{(m_{C_g^{p-}})_k} : L^1\big((m_{C_g^{p-}})_k\big) \to L^{r_k}(G)$$

is compact, for all $k \geqslant l$. Then [25, Theorem 1.2] yields that $g \in L^{r_k}(G)$, for all $k \geqslant l$, and since $L^s(G) \subseteq L^r(G)$ for $s > r$ we obtain that

$$g \in \bigcap_{k \geqslant l} L^{r_k}(G) = \bigcap_{k \in \mathbb{N}} L^{r_k}(G) = L^{p-}(G),$$

that is, condition (i) of Proposition 4.3.6 holds. $\quad\square$

Chapter 5

Conclusion

The results we have obtained in Chapter 3 revealed that the "optimal domain process" works also for continuous linear operators $T : X(\mu) \to X$ defined on a Fréchet function space $X(\mu)$ over a σ-finite measure space (Ω, Σ, μ) and with values in a Fréchet space X. Of course, as in the case of Banach function spaces, to obtain that the optimal domain of T is $L^1(m_T)$ and its optimal extension is the integration operator I_{m_T} it is necessary that the space $X(\mu)$ and the operator T fulfil certain requirements. First of all, the definition of the set function $m_T : \Sigma \to X$ associated with the operator T given by

$$m_T(A) := T(\chi_A), \quad \text{for } A \in \Sigma,$$

expects the space $X(\mu)$ to contain the Σ-simple functions $\text{sim}(\Sigma)$. Moreover, as in the case of Banach function spaces where the σ-order continuity is essential we cannot do without the σ-Lebesgue topology of $X(\mu)$ (which is the analogue to the σ-order continuity of a Banach function space). It ensures that $\text{sim}(\Sigma)$ is dense in $X(\mu)$, as established in Lemma 2.3.2, and that the finitely additive set function m_T becomes σ-additive and hence, a vector measure, as seen in Proposition 3.2.1. By Proposition 3.2.2 and Proposition 3.2.3 we know that under these conditions $X(\mu)$ is always contained in $L^1(m_T)$ and that the inclusion map $j_T : X(\mu) \to L^1(m_T)$ is continuous. However, to make sure that $X(\mu)$ is continuously *embedded* into $L^1(m_T)$ the inclusion map $j_T : X(\mu) \to L^1(m_T)$ needs to be injective; a property j_T has whenever the operator T is μ-determined (Proposition 3.2.4 and Lemma 3.2.2). Having all these "ingredients" available it follows that the Fréchet function space $L^1(m_T)$ is the largest amongst all Fréchet function spaces over (Ω, Σ, μ) having a σ-Lebesgue topology into which $X(\mu)$ is continuously embedded and to which T admits an X-valued continuous linear extension. Moreover, such an extension of T is unique and is precisely the integration operator $I_{m_T} : L^1(m_T) \to X$ (Theorem 3.3.1).

Such a strong result as Theorem 3.3.1 calls for applications. The operators we have chosen in Chapter 4 have received much attention when defined on a Banach function space. Therefore it was challenging to find out how the results differ when the operators were defined on a Fréchet function space. Since the Fréchet function spaces should contain the Σ-simple functions and have σ-Lebesgue topology the choice fell on the spaces $L^{p-}([0,1])$ resp. $L^{p-}(G)$ and $L^p_{\text{loc}}(\mathbb{R})$ (see Example 2.3.1 and Example 2.3.2).

The first results concern the *multiplication operator* $M_g^{p-} : L^{p-}([0,1]) \to L^{p-}([0,1])$, for $p \in (1, \infty)$ and $g \in \mathcal{M}^{p-}$ fixed, which is indeed a continuous linear operator (Proposition 4.1.1). Moreover, it is λ-determined (and thus, the theory of Chapter 3 is applicable) if and only if $g \neq 0$ λ-a.e. on $[0,1]$ (Proposition 4.1.2). It is then known that the optimal domain of M_g^{p-} is $L^1(m_{M_g^{p-}})$ and its optimal extension is $I_{m_{M_g^{p-}}}$. A characterization of $L^1(m_{M_g^{p-}})$ is given by

$$L^1(m_{M_g^{p-}}) = \left\{ f \in L^0([0,1]) : fg \in L^{p-}([0,1]) \right\}$$

(Proposition 4.1.3). The subsequent investigations showed that a major part of the results in Subsection 4.1.1 depends on the function g. A remarkable connection between the vector measure $m_{M_g^{p-}}$ and the spectral measure $\tilde{P} : \mathcal{B}([0,1]) \to L_s(L^{p-}([0,1]))$ given by $\tilde{P}(A) : f \mapsto f\chi_A$, for $A \in \mathcal{B}([0,1])$ and $f \in L^{p-}([0,1])$, however, allowed us to use the investigations in [1] to find a characterization of $\mathcal{M}^{p-}$, namely

$$\mathcal{M}^{p-} = \bigcap_{1 \leqslant s < \infty} L^s([0,1]).$$

Moreover, the same characterization of $\mathcal{M}^{p-}$ makes it possible to decide for which $g \in \mathcal{M}^{p-}$ the optimal domain coincides with $L^{p-}([0,1])$. This is done in Proposition 4.1.4 which states that

$$L^1(m_{M_g^{p-}}) = L^{p-}([0,1]) \text{ if and only if } \tfrac{1}{g} \in \mathcal{M}^{p-}.$$

Finally, further studies on the vector measure $m_{M_g^{p-}}$ showed that the variation of $m_{M_g^{p-}}$ is infinite for every $g \in \mathcal{M}^{p-} \setminus \{0\}$ (Proposition 4.1.5).

The investigation of the *multiplication operator* $M_{g,\text{loc}}^p : L^p_{\text{loc}}(\mathbb{R}) \to L^p_{\text{loc}}(\mathbb{R})$, where $p \in (1, \infty)$ and $g \in \mathcal{M}^p_{\text{loc}}$ is fixed, produced similar results. So, $M_{g,\text{loc}}^p$ turns out to be continuous (Proposition 4.1.6) and to be λ-determined if and only if $g \neq 0$ λ-a.e. on $\mathbb{R}$. In that case, the theory of Chapter 3 implies that the optimal domain of $M_{g,\text{loc}}^p$ is $L^1(m_{M_{g,\text{loc}}^p})$ and its optimal extension is the integral operator $I_{m_{M_{g,\text{loc}}^p}}$. A characterization of $L^1(m_{M_{g,\mathrm{loc}}^p})$ is

$$L^1(m_{M_{g,\mathrm{loc}}^p}) = \left\{ f \in L^0(\mathbb{R}) : fg \in L_{loc}^p(\mathbb{R}) \right\}$$

(Proposition 4.1.7). Again, there is a connection between the investigations in [1] concerning the spectral measure $\hat{P} : \mathcal{B}(\mathbb{R}) \to L_s(L_{loc}^p(\mathbb{R}))$ given by $\hat{P}(A) : f \mapsto f\chi_A$, for each $A \in \mathcal{B}(\mathbb{R})$ and $f \in L_{loc}^p(\mathbb{R})$, and the space $\mathcal{M}_{loc}^p$. Indeed, Proposition 4.1.8 identifies the space $\mathcal{M}_{loc}^p$ as

$$\mathcal{M}_{loc}^p = L_{loc}^\infty(\mathbb{R}) = L^1(\hat{P}).$$

The question whether $L^1(m_{M_{g,\mathrm{loc}}^p})$ is strictly larger than $L_{loc}^p(\mathbb{R})$ could not be answered without taking a closer look at the function g. So, whenever $\frac{1}{g} \in L_{loc}^\infty(\mathbb{R})$ the spaces $L^1(m_{M_{g,\mathrm{loc}}^p})$ and $L_{loc}^p(\mathbb{R})$ coincide, but as soon as $\frac{1}{g} \notin L_{loc}^\infty(\mathbb{R})$ this needs not to be the case and the inclusion $L_{loc}^p(\mathbb{R}) \subseteq L^1(m_{M_{g,\mathrm{loc}}^p})$ may indeed be proper. A last result on the vector measure $m_{M_{g,\mathrm{loc}}^p}$ associated with the operator $M_{g,\mathrm{loc}}^p$ concerns its variation. It is infinite, for every $g \in \mathcal{M}_{loc}^p \backslash \{0\}$ (Proposition 4.1.9). Note that in all these investigations the case $p = 1$ has not been considered.

The study of the *Volterra operator* $V_{p-} : L^{p-}([0,1]) \to L^{p-}([0,1])$, for $p \in (1,\infty)$, revealed that many of the results obtained in Section 4.2 resemble those established in [28]; there the Volterra operator V_r defined on the Banach function space $L^r([0,1])$, for $1 < r < \infty$, was investigated. First of all, since V_{p-} is continuous (Proposition 4.2.1) and injective (Proposition 4.2.2), thus, by Corollary 3.2.1 also λ-determined, we can apply the theory of Chapter 3 to conclude that $L^1(m_{V_{p-}})$ is the optimal domain of V_{p-} and its optimal extension the integral operator $I_{m_{V_{p-}}}$. However, the investigations should not only concentrate on the space $L^1(m_{V_{p-}})$ but also include the spaces $\bigcap_{k \in \mathbb{N}} L^1(|(m_{V_{p-}})_k|)$ and $\bigcap_{k \in \mathbb{N}} L^1((m_{V_{p-}})_k)$. With the aid of the function $g_{p-} : [0,1] \to L^{p-}([0,1])$ given by

$$g_{p-}(t) := \chi_{[t,1]}, \quad \text{for } t \in [0,1],$$

it is possible to give a full characterization of the spaces $\bigcap_{k \in \mathbb{N}} L^1(|(m_{V_{p-}})_k|)$ and $L^1(m_{V_{p-}})$ (Lemma 4.2.2 and Lemma 4.2.3). In summary, concerning the connections between the different spaces, we have the following inclusions:

$$L^{p-}([0,1]) \subsetneqq L^1([0,1]) \subsetneqq L^1(m_{V_{p-}}),$$

yielding that the optimal domain of V_{p-} is indeed strictly larger than $L^{p-}([0,1])$ (Proposition 4.2.4 and Proposition 4.2.5). Moreover, Proposition 4.2.6 established the equality

$$L^1(m_{V_{p-}}) = \bigcap_{k \in \mathbb{N}} L^1\big((m_{V_{p-}})_k\big).$$

Still open, however, is the question whether the inclusion

$$\bigcap_{k \in \mathbb{N}} L^1\big(|(m_{V_{p-}})_k|\big) \subseteq L^1(m_{V_{p-}})$$

is strict (which would be the respective result to the case of V_r defined on the Banach function spaces $L^r([0,1])$, where $1 < r < \infty$) or reduces to an equality. Further investigations took a closer look at the variation of $m_{V_{p-}}$ and the properties of the integration operator $I_{m_{V_{p-}}}$. Since $m_{V_{p-}}$ coincides with the Bochner λ-integral of g_{p-} (see the discussion following Lemma 4.2.1) its variation is finite. And as in the case of the Volterra operator V_r defined on the Banach function space $L^r([0,1])$, where $1 < r < \infty$, the integration operator $I_{m_{V_{p-}}}$ is not compact (Proposition 4.2.7).

Regarding the *convolution operator* $C_g^{p-} : L^{p-}(G) \to L^{p-}(G)$, where $p \in (1,\infty)$ and G a compact Abelian group, we obtained a couple of results which keep on the whole to the results established in [25]. Since C_g^{p-} is continuous (Proposition 4.3.1) and μ-determined, for each $g \in L^1(G)\backslash\{0\}$ (Proposition 4.3.2), it is clear from the theory of Chapter 3 that the optimal domain of C_g^{p-} is $L^1(m_{C_g^{p-}})$ and its optimal extension is the integration operator $I_{m_{C_g^{p-}}}$. A first characterization of the space $L^1(m_{C_g^{p-}})$ shows that

$$L^1(m_{C_g^{p-}}) = \big\{ f \in L^1(G) : (f\chi_A) * g \in L^{p-}(G), \text{ for all } A \in \mathcal{B}(G) \big\},$$

whenever $g \in L^1(G)\backslash\{0\}$ satisfies $g \geqslant 0$ (Proposition 4.3.3). A more interesting result can be achieved when g is an element of the smaller space $L^{p-}(G)$. By making use of the continuous and Bochner μ-integrable function $F_g^{p-} : G \to L^{p-}(G)$ given by

$$F_g^{p-}(y) := g(\cdot - y), \quad \text{for } y \in G$$

(Proposition 4.3.4 and Proposition 4.3.5), it is possible to establish seven equivalent assertions which hold if and only if $g \in L^{p-}(G)\backslash\{0\}$ (Proposition 4.3.6 and Proposition 4.3.7). Among others we obtain the equality $L^1(m_{C_g^{p-}}) = L^1(G)$. Since $m_{C_g^{p-}}$ coincides in that case with the Bochner μ-integral of F_g^{p-} it follows that $m_{C_g^{p-}}$ is of finite variation. And last but not least, the integration operator $I_{m_{C_g^{p-}}}$ turns out to be compact if and only if $g \in L^{p-}(G)$.

Nevertheless, the operators investigated in Chapter 4 represent only a small part of the applications that are possible with the results established in this thesis. Many more operators T defined on Fréchet function spaces (others than only $L^{p-}([0,1])$, $L^{p-}(G)$ and $L^p_{\text{loc}}(\mathbb{R})$) wait to be studied. Prospective investigations will not only be interesting in view of the optimal domain $L^1(m_T)$ of those operators and the question whether $L^1(m_T)$ is strictly larger than the original domain, but also in view of their optimal extension, the integration operator I_{m_T}, and its properties.

Bibliography

[1] J. Bonet, W. J. Ricker: *Spectral measures in classes of Fréchet spaces.* Bulletin de la Société Royale des Sciences de Liège 73, 99–117 (2004).

[2] J. M. F. Castillo, J. C. Díaz, J. Motos: *On the Fréchet space L_{p-}.* Manuscripta Math. 96, 219–230 (1998).

[3] G. P. Curbera, W. J. Ricker: *Optimal domains for kernel operators via interpolation.* Math. Nachr. 244, 47–63 (2002).

[4] G. P. Curbera, W. J. Ricker: *Optimal domains for the kernel operator associated with Sobolev's inequality.* Studia Math. 158, 131–152 (2003), and 170, 217–218 (2005).

[5] R. del Campo, W. J. Ricker: *The space of scalarly integrable functions for a Fréchet-space-valued measure.* J. Math. Anal. Appl. 354, 641–647 (2009).

[6] R. del Campo, W. J. Ricker: *Notes on scalarly integrable functions and Fréchet function spaces.* Unpublished manuscript.

[7] R. del Campo, W. J. Ricker: *The Fatou completion of a Fréchet function space and applications.* J. Aust. Math. Soc. 88, 49–60 (2010).

[8] J. Diestel, J. J. Uhl Jr.: *Vector Measures.* Mathematical Surveys, Vol. 15. Amer. Math. Soc., Providence, RI (1977).

[9] N. Dinculeanu: *Vector Measures.* Hochschulbücher für Mathematik, Vol. 64. Dt. Verl. der Wiss., Berlin (1966).

[10] N. Dunford, J. T. Schwartz: *Linear Operators I: General Theory.* 2nd printing. Wiley-Interscience Publ., New York (1964).

[11] J. Elstrodt: *Maß- und Integrationstheorie.* 6th revised edition. Springer-Verlag, Berlin-Heidelberg (2009).

[12] D. H. Fremlin: *Measure Theory.* Vol. 2: Broad Foundations. Torres Fremlin, Colchester (2003).

[13] I. M. Gelfand, G. E. Shilov: *Generalized Functions*. Vol. 2: Spaces of Fundamental and Generalized Functions. Academic Press, New York (1968).

[14] E. Hewitt, K. A. Ross: *Abstract Harmonic Analysis I*. Grundlehren der mathematischen Wissenschaften, Vol. 115. Springer-Verlag, Berlin (1963).

[15] E. Hewitt, K. A. Ross: *Abstract Harmonic Analysis II*. Grundlehren der mathematischen Wissenschaften, Vol. 152. Springer-Verlag, Berlin (1970).

[16] V. M. Klimkin, M. G. Svistula: *Darboux property of a non-additive set function.* SB MATH 192(7), 969–978 (2001).

[17] I. Kluvánek, G. Knowles: *Vector Measures and Control Systems*. North-Holland, Amsterdam (1971).

[18] G. Köthe: *Topological Vector Spaces I*. Grundlehren der mathematischen Wissenschaften, Vol. 159. 2nd revised edition. Springer, Berlin (1983).

[19] D. R. Lewis: *Integration with respect to vector measures*. Pacific J. Math. 33, 157–165 (1970).

[20] D. R. Lewis: *On integrability and summability in vector spaces*. Illinois J. Math. 16, 294–307 (1972).

[21] C. W. McArthur: *On a theorem of Orlicz and Pettis*. Pacific J. Math. 22, 297–302 (1967).

[22] R. Meise, D. Vogt: *Einführung in die Funktionalanalysis*. Vieweg-Studium, Vol. 62. Vieweg, Braunschweig-Wiesbaden (1992).

[23] R. Meise, D. Vogt: *Introduction to Functional Analysis*. Oxford Graduate Texts in Mathematics, Vol. 2. Clarenden Press, Oxford (1997).

[24] S. Okada, W. J. Ricker: *Compact integration operators for Fréchet-space-valued measures*. Indag. Mathem., N.S., 13, 209–227 (2002).

[25] S. Okada, W. J. Ricker: *Optimal domains and integral representations of convolution operators in $L^p(G)$*. Integr. Equ. Oper. Theory 48, 525–546 (2004).

[26] S. Okada, W. J. Ricker, E. A. Sánchez Pérez: *Optimal Domain and Integral Extension of Operators Acting in Function Spaces*. Operator Theory Advances and Applications, Vol. 180. Birkhäuser, Basel (2008).

[27] Boto von Querenburg: *Mengentheoretische Topologie*. 3rd revised edition. Springer, Berlin (2001).

[28] W. J. Ricker: *Compactness properties of extended Volterra operators in $L^p([0,1])$ for $1 \leqslant p \leqslant \infty$.* Archiv der Mathematik 66, 132–140 (1996).

[29] W. J. Ricker: *Operator Algebras Generated by Commuting Projections: a Vector Measure Approach.* Lecture Notes in Mathematics, Vol. 1711. Springer, Berlin et al. (1999).

[30] W. Rudin: *Fourier Analysis on Groups.* Interscience Tracts in Pure and Applied Mathematics, Vol. 12. Interscience Publ., New York (1962).

[31] W. Rudin: *Reelle und komplexe Analysis.* Oldenburg, München-Wien (1999).

[32] H. H. Schaefer: *Topological Vector Spaces.* Graduate Texts in Mathematics, Vol. 3. 2nd edition. Springer, New York (1999).

[33] A. E. Taylor: *General Theory of Functions and Integration.* Blaisdell Publ. Co., New York (1965).

[34] G. E. R. Thomas: *Integration of functions with values in locally convex Suslin spaces.* Trans. Amer. Math. Soc. 212, 61–81 (1975).

[35] D. Werner: *Funktionalanalysis.* 6th revised edition. Springer, New York-Heidelberg-Berlin (2007).

[36] A. C. Zaanen: *Integration.* 2nd edition. North-Holland, Amsterdam (1967).

Index

group, dual, 51

group, locally compact, 49

group, topological, 48

Hölder's inequality, 18

Haar integral, 49

Haar measure, 49

Hausdorff space, 10, 11

ideal, 28

indefinite integral, 40

integration operator, 69

joint Riesz-Fischer property, 29

Lebesgue's dominated convergence theorem, 24

local Banach space, 14, 40

local convergence in measure, 19

locally compact group, 49

locally compact topological vector space, 15

locally convex topological vector space, 10

locally solid Riesz space, 28

measurable space, 15

measure, 16

measure space, 16

measure space, complete, 16

measure, σ-finite, 16

measure, complex, 25

measure, control, 44, 68

measure, finite, 16

measure, non-atomic, 26

measure, spectral, 80

measure, variation, 25

metric, 12

metrizability, 12

metrizable function space, 28

Minkowski functional, 11

modulus, 27

monotone convergence theorem, 24

multiplication operator, 75, 86

neighbourhood, 10

neighbourhood base, 10

non-atomic measure, 26

norm, 10

operator, convolution, 112

operator, multiplication, 75, 86

operator, translation, 119

operator, Volterra, 95

Orlicz-Pettis theorem, 14

Pettis μ-integrable, 48

polar set, 41

positive cone, 27, 28

pseudo-metric, 12

reflexive, 14

Riesz space, 27

Riesz space, locally solid, 28

scalarly m-integrable, 39

semi-norm, 10

separated, 11

sequential completeness, 11

solid, 28

spectral measure, 80

strong topology, 14

strongly μ-measurable, 47

subseries convergence, 14

topological group, 48

topological vector space, 10

topological vector space, locally compact, 15

topological vector space, locally convex, 10

topology, 9

topology of local convergence, 20